Maths

Foundation Tier

Mike Fawcett

Practice Test Papers

Contents

PRACTICE EXAM 1

PRACTICE EXAM 2

ANSWERS

ACKNOWLEDGEMENTS

The author and publisher are grateful to the copyright holders for permission to use quoted materials and images.

Cover and page 1: © Shutterstock.com/homoparadoksuma

Every effort has been made to trace copyright holders and obtain their permission for the use of copyright material. The author and publisher will gladly receive information enabling them to rectify any error or omission in subsequent editions. All facts are correct at time of going to press.

Published by Letts Educational
An imprint of HarperCollins*Publishers*
1 London Bridge Street
London SE1 9GF

ISBN: 9780008166694

First published 2016

10 9 8 7 6 5 4 3 2 1

British Library Cataloguing in Publication Data.

A CIP record of this book is available from the British Library.

Commissioning Editor: Emily Linnett
Author: Mike Fawcett
Project Management: Richard Toms
Cover Design: Paul Oates
Inside Concept Design: Ian Wrigley
Text Design and Layout: Contentra Technologies
Production: Lyndsey Rogers
Printed in China by RR Donnelley APS

Letts

GCSE

Mathematics

F

Foundation tier

Paper 1 Time: 1 hour 30 minutes

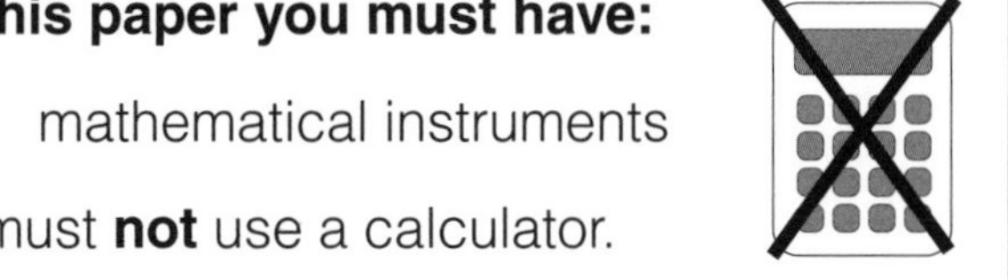

Instructions

- Use black ink or black ball-point pen. Draw diagrams in pencil.
- Read each question carefully before you start to write your answer.
- Diagrams are **not** accurately drawn unless otherwise stated.
- Answer **all** questions.
- You must answer the questions in the space provided.
- In all calculations, show clearly how you work out your answer. Use a separate sheet of paper if needed. Marks may be given for a correct method even if the answer is wrong.

Information

- The marks for each question are shown in brackets.
- The maximum mark for this paper is 80.

Name: ....

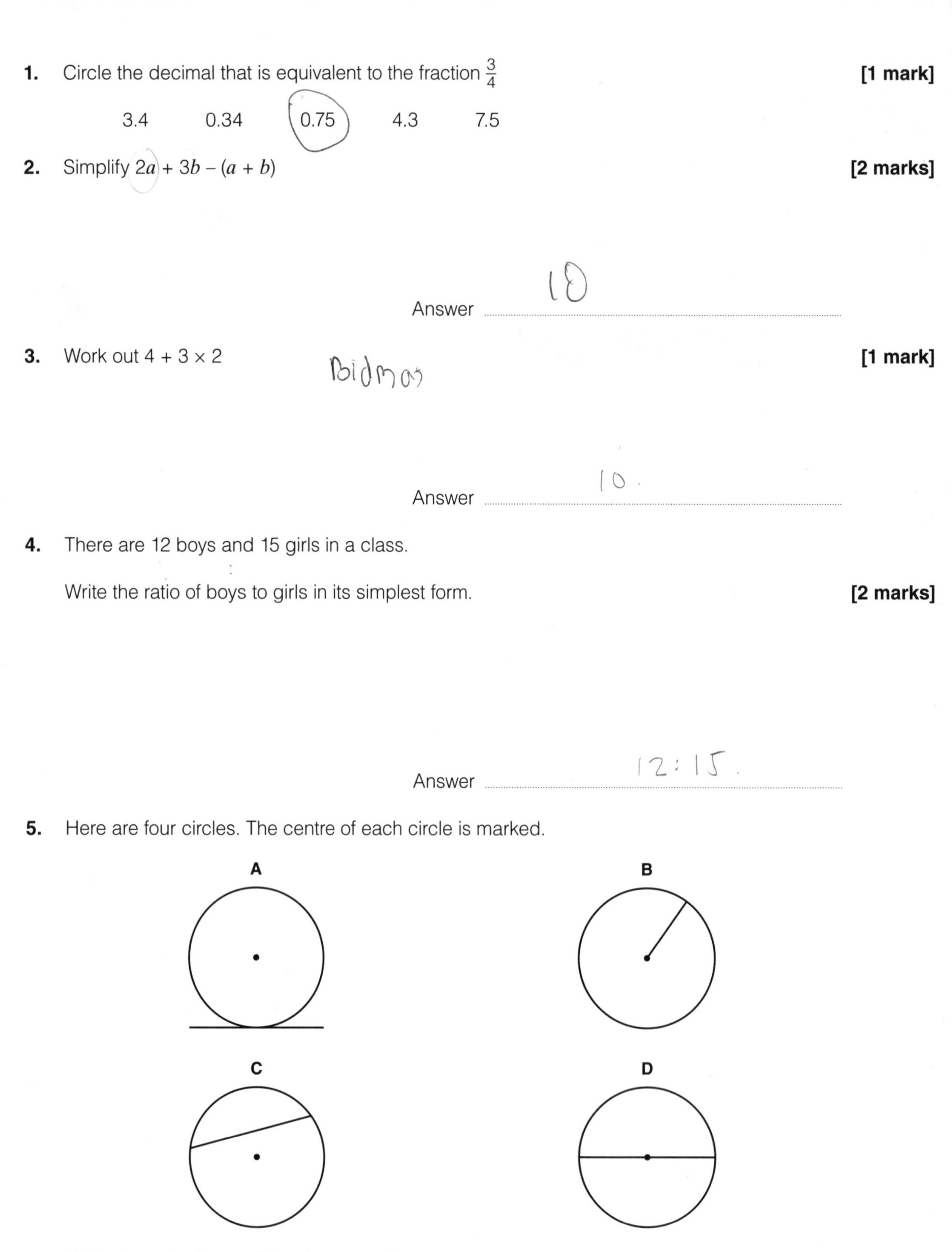

1. Circle the decimal that is equivalent to the fraction $\frac{3}{4}$ **[1 mark]**

3.4 0.34 0.75 4.3 7.5

2. Simplify $2a + 3b - (a + b)$ **[2 marks]**

Answer ....

3. Work out $4 + 3 \times 2$ **[1 mark]**

Answer ....

4. There are 12 boys and 15 girls in a class.

Write the ratio of boys to girls in its simplest form. **[2 marks]**

Answer ....

5. Here are four circles. The centre of each circle is marked.

Write down the letter of the diagram which shows:

(a) A radius Answer .... **[1 mark]**

(b) A diameter Answer .... **[1 mark]**

6. There are 20 sweets in a bag.

Seven of them are strawberry.

A sweet is taken at random from the bag.

What is the probability that the sweet taken is **not** strawberry? **[2 marks]**

Answer ....

7. Give an example to show that each of these statements is incorrect.

(a) A prime number is always odd. **[1 mark]**

Answer ....

(b) The factors of 16 are always even. **[1 mark]**

Answer ....

(c) The difference between two square numbers is always odd. **[1 mark]**

Answer ....

8. The following four cards have a median of 16 and a range of 8.

18	12	?	14

Find the value of the missing card. **[2 marks]**

Answer ....

9. The sum of three numbers is 16.

Each of the three numbers is a factor of 24.

What could the three numbers be? **[4 marks]**

Answer ...., ... and ....

10. Solve the equation $4(x - 2) = 2$ **[2 marks]**

Answer $x =$ ....

11. Mike takes his two children, Bethany and Ethan, to a BMX track.

All three are going to take part.

Here are the costs for the BMX session:

	Child	**Adult**
BMX session	£6.40	£8.25
Bike hire	£2.20	£3.50
Helmet hire	£1.50	£2.00
Pads hire	£1.50	£2.00

Ethan takes his own bike, helmet and pads.

Bethany takes her own bike, **but** needs to hire helmet and pads.

Mike needs to hire a bike, helmet and pads.

Find the total cost of the BMX trip. **[3 marks]**

Answer £ ....

12. Melanie has chosen to do three activities at an outdoor adventure park.

They are:

- Air Jump
- 3G Swing
- High Ropes

She can do them in any order she likes.

How many different possible combinations are there? **[3 marks]**

Answer ....

13. Here are the first four patterns in a sequence.

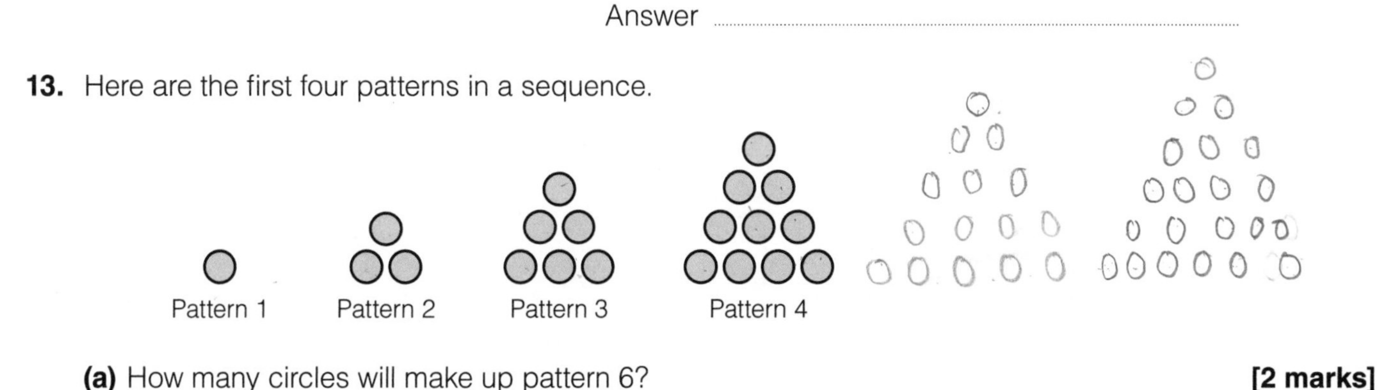

(a) How many circles will make up pattern 6? **[2 marks]**

Answer ....

(b) Describe the sequence using the words 'Odd numbers' and 'Even numbers'. **[1 mark]**

....

....

14. Draw the graph of $y = 2x - 1$ for values of x from –2 to 3. **[3 marks]**

15. A school had 1200 students in 2014. In 2015 the number of students had increased by 15%.

(a) How many students attended the school in 2015? **[2 marks]**

Answer .... students

From 2015 to 2016, the number of students decreased by 10%.

(b) How many students attended the school in 2016? **[2 marks]**

Answer .... students

16. Shop A sells a pack of five avocados for £1.40

Shop B sells a pack of four avocados for £1.16

Which shop gives the best value for money?

You must show all of your working. **[3 marks]**

5 x 1.40 =

4 x 1.16 =

Answer ....

17. Tiles are to be placed along a section of a kitchen wall as shown.

The wall is 1.4 m long. The tiles are 250 mm wide.

(a) How many full tiles can fit across the length of wall? **[2 marks]**

Answer .... tiles

(b) What length of the wall would be left uncovered? Give your answer in cm. **[2 marks]**

Answer .... cm

18. Work out $\frac{3}{5} + \frac{2}{7}$ **[2 marks]**

$\frac{3}{5} + \frac{2}{7} = 5$

Answer ....

19. Christopher and Liam both have a photograph printed.

The two photographs are similar rectangles.

The length of Liam's photograph is half the length of Christopher's photograph.

Christopher says, "The area of my photograph is double the area of your photograph."

Is Christopher correct?

Show all your working. **[4 marks]**

20. Charlotte, Clarisse and Mathilda share some money in the ratio 4 : 6 : 11

Mathilda receives £200 more than Clarisse.

How much do they each receive? **[3 marks]**

Charlotte £ ....

Clarisse £ ....

Mathilda £ ....

21. AE, BF and DCG are parallel.

$BC = CD$

$ADB = 22°$

$EAB = 68°$

Work out the size of the angle marked x. You must show your working. **[4 marks]**

Answer .... °

22. Tim and Amanda want to investigate the probability that it will rain on any given day in October.

Tim records the number of days that it rains in a week. Amanda records the number of days that it rains in the whole month.

Here are their results:

	Number of days it rains	**Number of days without rain**
Tim's results	2	5
Amanda's results	11	20

(a) Write down two different estimates for the probability that it will rain on any given day in October. **[2 marks]**

Answer .... and ....

(b) Which is the most reliable estimate from your answers in part (a)?

Give a reason for your answer. **[2 marks]**

Answer ....

Reason ....

....

23. **(a)** Solve the inequality $7x - 4 \leqslant 7 + 5x$ **[2 marks]**

Answer ....

(b) Show your answer to part (a) on the number line. **[2 marks]**

0 1 2 3 4 5 6 7 8 9 10

24. Work out the answer to $(6.1 \times 10^5) + (4.2 \times 10^4)$

Give your answer in standard form. **[2 marks]**

Answer ....

25. Expand and simplify $(2x + 3)(4x - 2)$ **[2 marks]**

Answer ....

26. A house was bought in 2012. It was sold in 2016 for £180 000. The seller made 20% profit.

How much did the seller buy the house for in 2012? **[2 marks]**

Answer £ ....

27. BC is parallel to DE.

$BC = 4$ cm

$DE = 10$ cm

$AD = 7.5$ cm

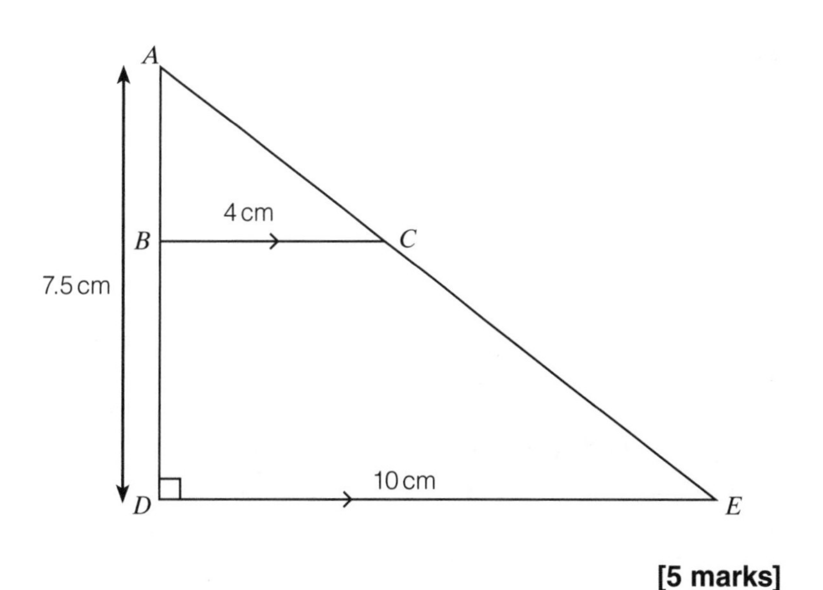

Work out the length of CE. **[5 marks]**

Answer .... cm

28. Mr Chan's tutor group has 27 students. They each decide what event they will take part in on sports day.

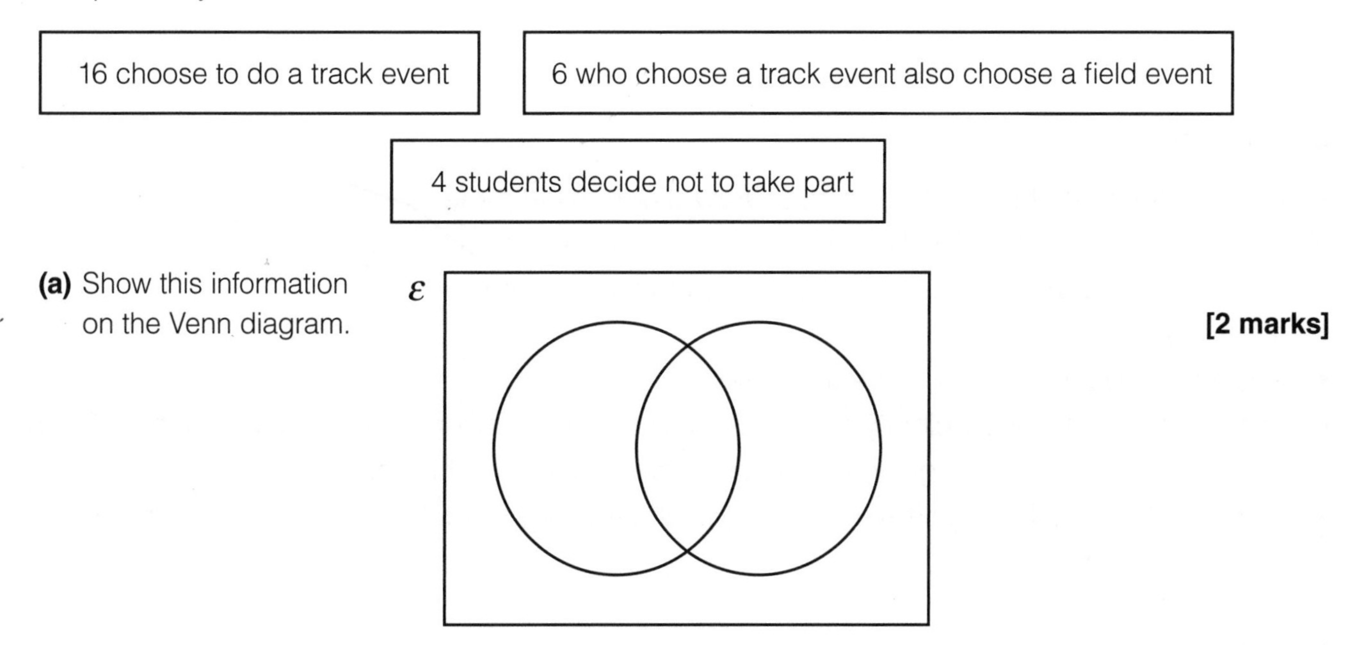

(a) Show this information on the Venn diagram. **[2 marks]**

(b) A student is chosen at random from the tutor group.

Find the probability that this student decided to do both a track event and a field event. **[2 marks]**

Answer ....

END OF QUESTIONS

GCSE

Mathematics

F

Foundation tier

Paper 2

Time: 1 hour 30 minutes

For this paper you must have:

- a calculator
- mathematical instruments

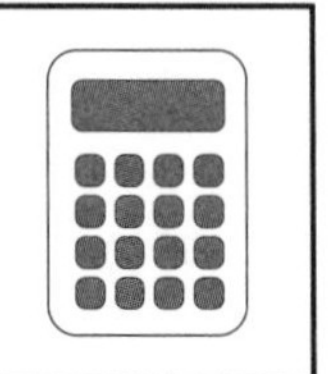

Instructions

- Use black ink or black ball-point pen. Draw diagrams in pencil.
- Read each question carefully before you start to write your answer.
- Diagrams are **not** accurately drawn unless otherwise stated.
- Answer **all** the questions.
- Answer the questions in the space provided.
- In all calculations, show clearly how you work out your answer. Use a separate sheet of paper if needed. Marks may be given for a correct method even if the answer is wrong.
- If your calculator does not have a π button, take the value of π to be 3.142 unless the question instructs otherwise.

Information

- The marks for each question are shown in brackets.
- The maximum mark for this paper is 80.

Name: ....

1. **(a)** Write down 20% of £60. **[1 mark]**

Answer £ ....

(b) Write 40% as a fraction in its simplest form. **[1 mark]**

Answer ....

(c) Write 38% as a decimal. **[1 mark]**

Answer ....

2. This diagram is drawn accurately.

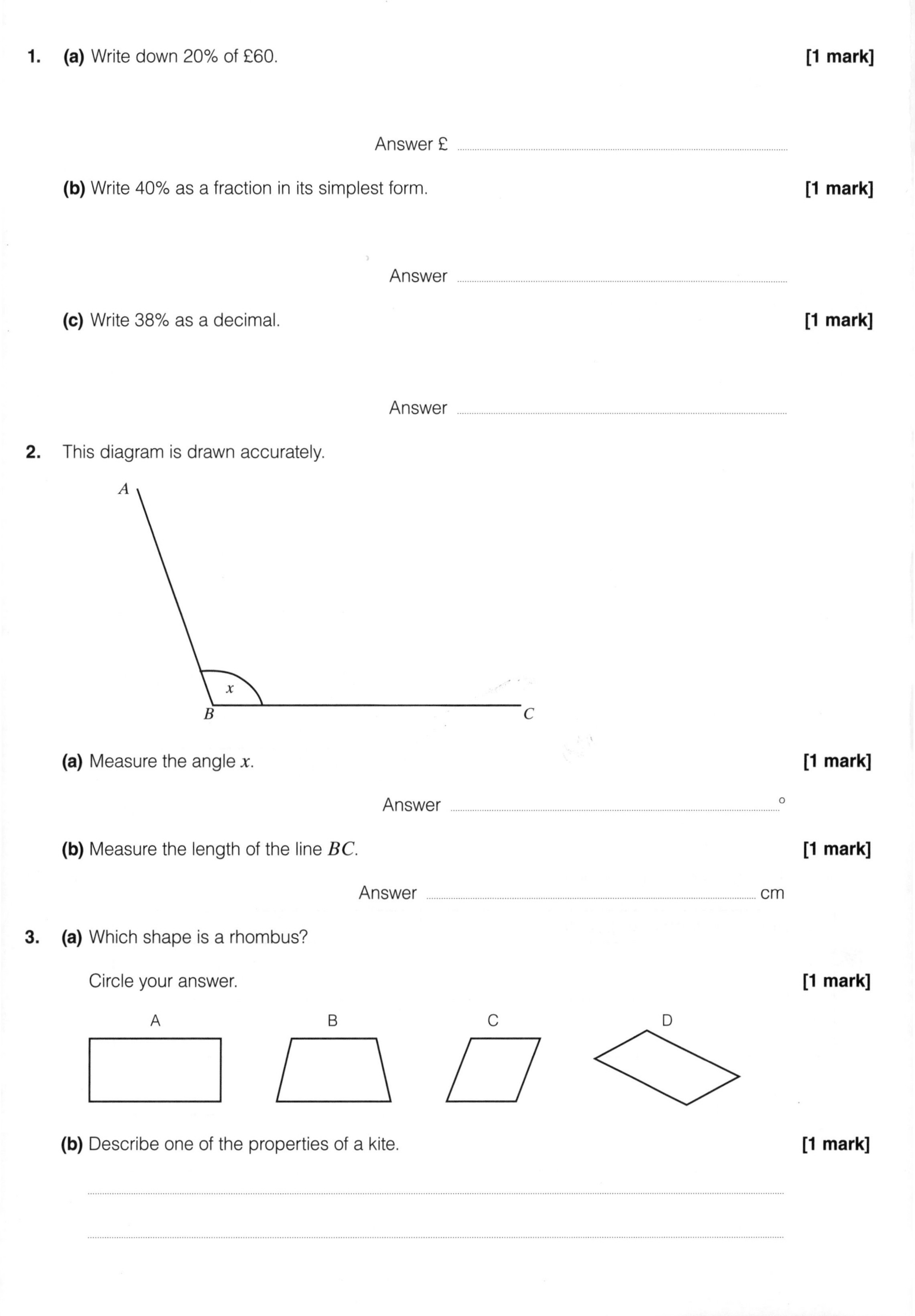

(a) Measure the angle x. **[1 mark]**

Answer ....°

(b) Measure the length of the line BC. **[1 mark]**

Answer .... cm

3. **(a)** Which shape is a rhombus?

Circle your answer. **[1 mark]**

A B C D

(b) Describe one of the properties of a kite. **[1 mark]**

....

....

4. There are six tiles in a bag, each with a letter on.

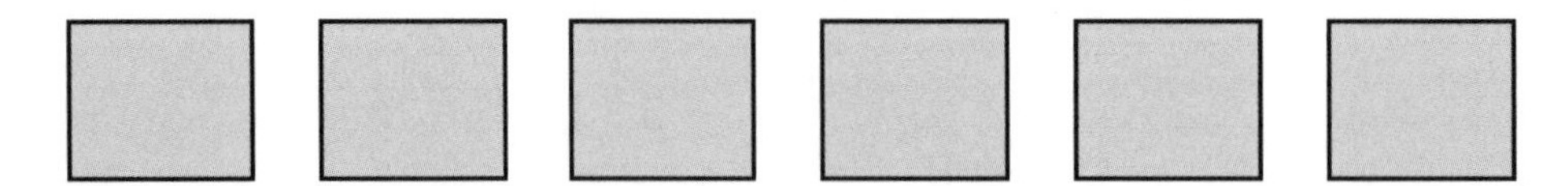

The probability of randomly choosing a tile labelled A, a tile labelled B or a tile labelled C is marked on the probability scale.

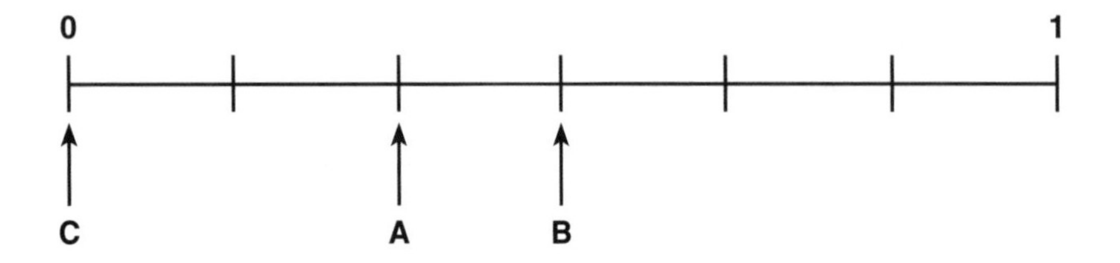

Write down a letter on each of the six tiles to make the probability scale true. **[2 marks]**

5. Write down whether each of these statements is **true** or **false**.

(a) $-3 < -1$ Answer .... **[1 mark]**

(b) $0.07 > 0.6$ Answer .... **[1 mark]**

(c) $3^2 = 6$ Answer .... **[1 mark]**

(d) $8^1 = 8$ Answer .... **[1 mark]**

6. The table shows the activities which some people took part in at a leisure centre.

Activity	Number of people
Swimming	18
Gym	15
Badminton	8
Squash	10
Other	4

Show this information in a pictogram. **[3 marks]**

Activity	Number of people
Swimming	
Gym	
Badminton	
Squash	
Other	

Key:

7. Here is a number machine.

(a) When the input is 4, the output is 3.

Write down the possible values of A and B. **[1 mark]**

Answer A = ... and B = ...

(b) Work out the input when the output is 4.5 **[2 marks]**

Answer ....

8. Richard buys x boxes of erasers for his class. There are y erasers in each box.

Richard already had 7 erasers.

(a) Write down an expression in terms of x and y to show how many erasers he now has. **[2 marks]**

Answer ....

Richard now has 39 erasers. There were 8 in each box.

(b) How many boxes did Richard buy? **[3 marks]**

Answer .... boxes

9. Write down the nth term for the following sequence.

5, 8, 11, 14, 17, … **[2 marks]**

Answer ....

10. The map shows the area around Kinder Scout, which is in the Peak District.

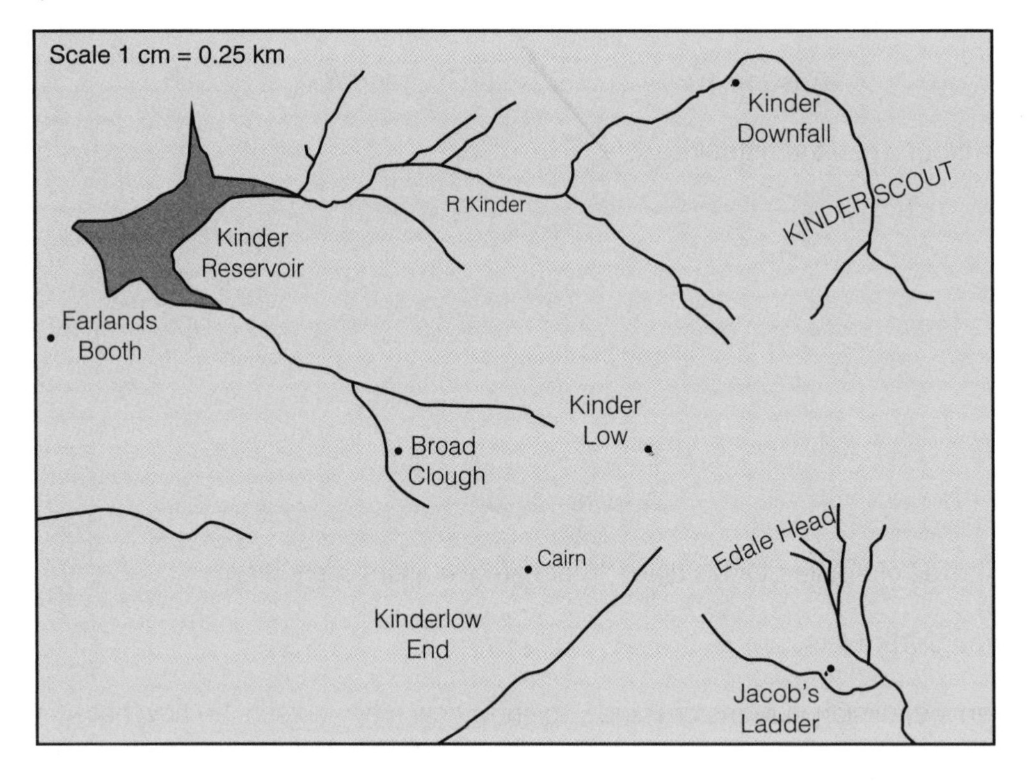

Lee and Ramesh are at Kinder Downfall. They are going to walk to Kinder Low and then from Kinder Low to Jacob's Ladder.

(a) Work out the actual distance that they will walk. **[2 marks]**

Answer .... km

The journey takes Lee and Ramesh 45 minutes.

(b) Work out their average speed. **[3 marks]**

Answer .... km/h

11. The price to hire a snooker table is £3 per hour plus a fixed charge of £1.50

(a) Use this information to plot a graph of cost against time. **[3 marks]**

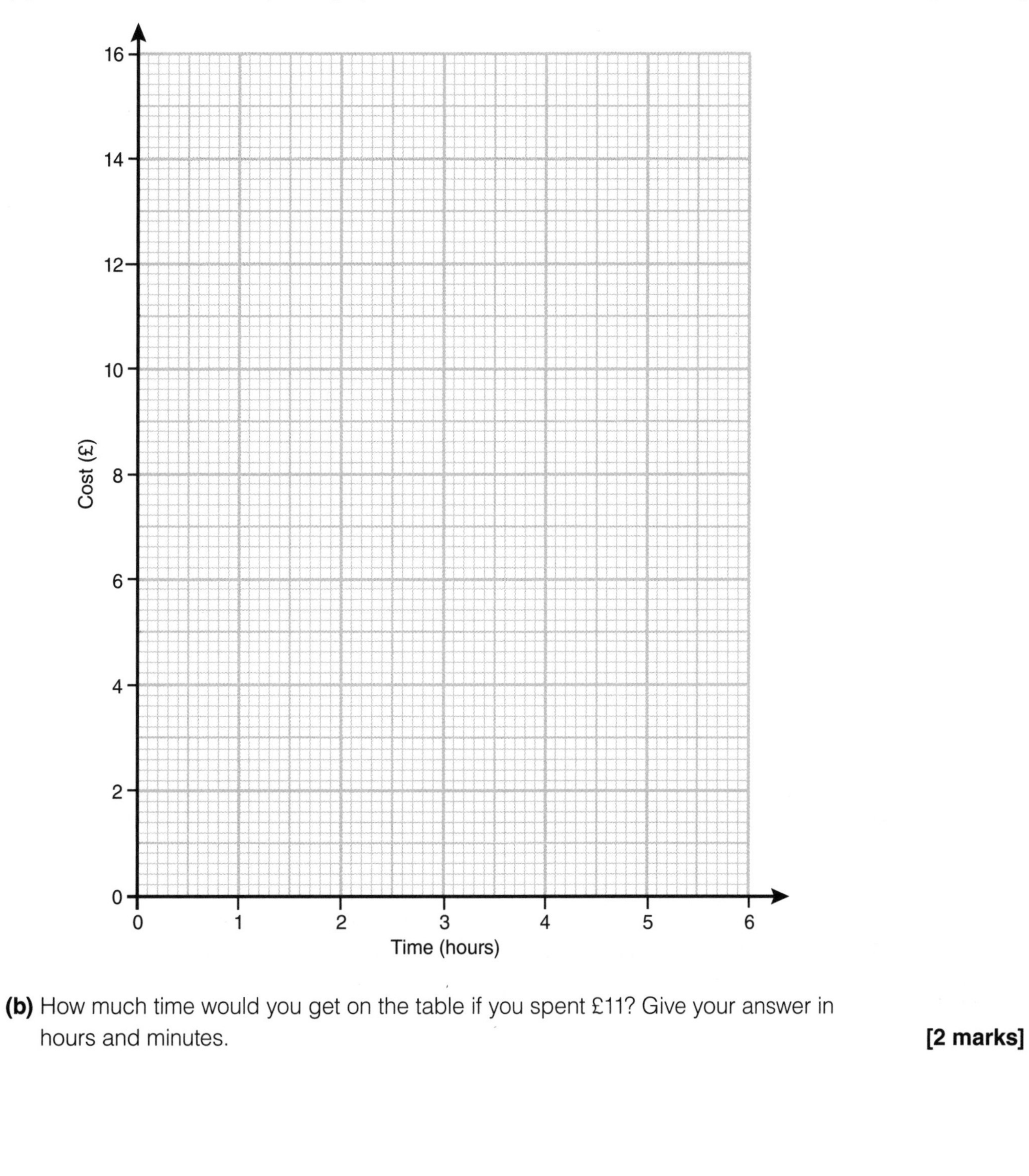

(b) How much time would you get on the table if you spent £11? Give your answer in hours and minutes. **[2 marks]**

Answer .... hours and .... minutes

12. Rearrange to make a the subject.

$v = u + at$ **[2 marks]**

Answer ....

13. Rachel (R), Jo (J) and Anuja (A) are competing in a running race.

Assume that there are no joint places and that each runner has an equal chance of winning.

(a) List all the different orders in which they could finish the race.

The first one is done for you. **[2 marks]**

1st	2nd	3rd
R	J	A

(b) What is the probability that Jo will finish in a slower time than Rachel? **[2 marks]**

Answer ....

14. The three diagrams show the plan view, front elevation and side elevation of a 3D shape.

Front elevation

Side elevation

Plan view

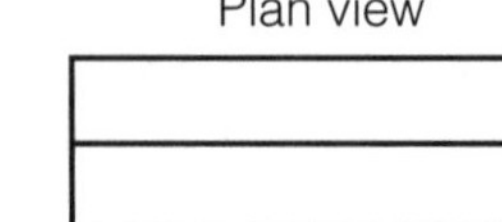

Sketch the 3D shape below. **[2 marks]**

15. The table shows the salary of five midwives and the number of years that they have been working.

Years	2	5	3	11	8
Salary	£22500	£28000	£24000	£31500	£30000

(a) Draw a scatter graph to represent this information. **[2 marks]**

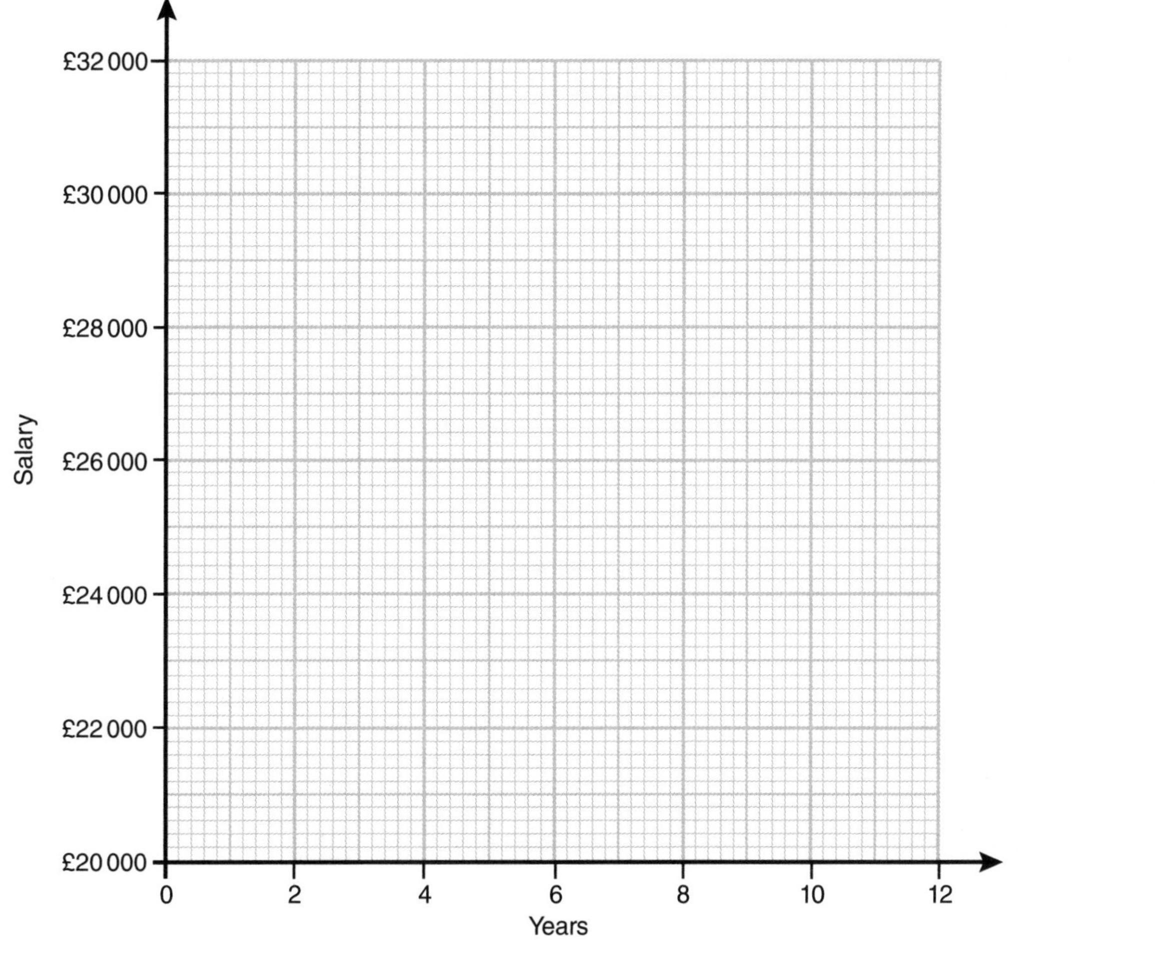

(b) Estimate the salary of a midwife who has been working for six years. **[1 mark]**

Answer £ ....

(c) Explain why the graph would not be suitable to estimate the salary of a midwife who has been working for 20 years. **[1 mark]**

....

....

16. This shot-put landing area is in the shape of a sector.

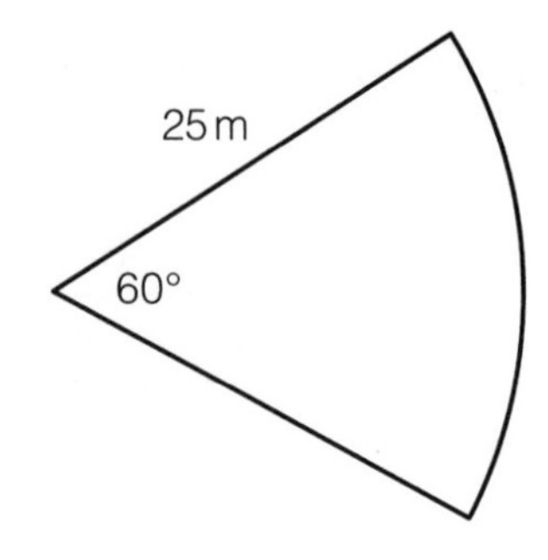

Find the length of the perimeter of the shot-put landing area. Give your answer to 3 significant figures. **[4 marks]**

Answer .... m

17. (a) Factorise fully $12x^2 - 4x$

Circle your answer. **[1 mark]**

$12x(x - 4)$ $4(3x^2 - x)$ $2x(6x - 2)$ $4x(3x - 1)$

(b) Simplify $3x^2y^{-1} \times 4xy^3$

Circle your answer. **[1 mark]**

$12x^3y^2$ $7x^3y^4$ $12xy^4$ $7x^3y^{-3}$

(c) Substitute $x = -2$ into the expression $3x^2 - x$

Circle your answer. **[1 mark]**

34 38 14 10

18. **(a)** Use approximation to estimate the answer to $\frac{3.9^3}{\sqrt{96.8} - 6.16}$ **[2 marks]**

Answer ....

(b) Use your calculator to write down the exact answer to part (a).

Write down all the figures on your calculator display. **[1 mark]**

Answer ....

(c) Round your answer in part (b) to 3 significant figures. **[1 mark]**

Answer ....

19. Melanie changes £450 into euros (€) to take on a trip to France.

The exchange rate is £1 = €1.285

(a) Work out how many euros Melanie should get. **[2 marks]**

Answer € ....

Melanie sees a bottle of perfume in France. It costs €36. She knows that the same bottle of perfume costs £27 in England.

(b) What is the difference in price between the two countries? Give your answer in pounds. **[2 marks]**

Answer £ ....

20. Stu is buying tins to make up some food bags for a food bank. Tins of beans are sold in boxes of 14. Tins of soup are sold in boxes of 8.

He wants to buy enough tins of beans and soup so that he can put one of each in each bag.

How many boxes of beans and soup should he buy so that he has none left over? **[3 marks]**

Answer .... boxes of beans and .... boxes of soup

21. Dave is making a concrete mix. He mixes together 500 g of cement, 1 kg of sand and 1.5 kg of water. He makes 2.17 litres of concrete.

Work out the density of the concrete. Give your answer to 2 decimal places. **[2 marks]**

Answer .... g/cm^3

22. A house was valued at £152 000 in 2013. The following year the value increased by 2%.

In the two years that followed, the value increased at a rate of 2.5% per year.

How much was the house worth in 2016? **[3 marks]**

Answer £ ....

23. **(a)** Evaluate 13^0 **[1 mark]**

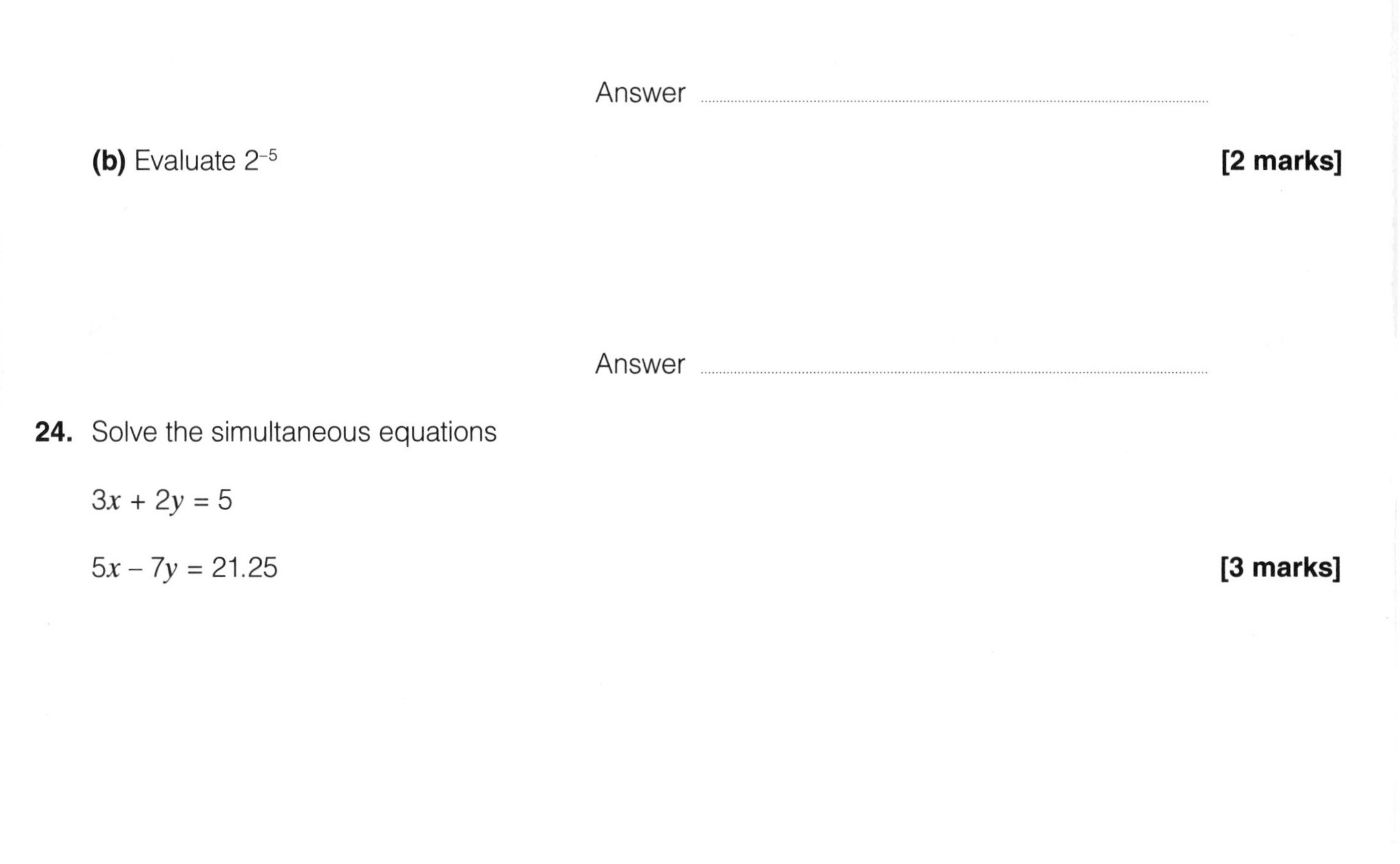

Answer ....

(b) Evaluate 2^{-5} **[2 marks]**

Answer ....

24. Solve the simultaneous equations

$3x + 2y = 5$

$5x - 7y = 21.25$ **[3 marks]**

Answer $x =$ .... and $y =$ ....

25. RST is a right-angled triangle.

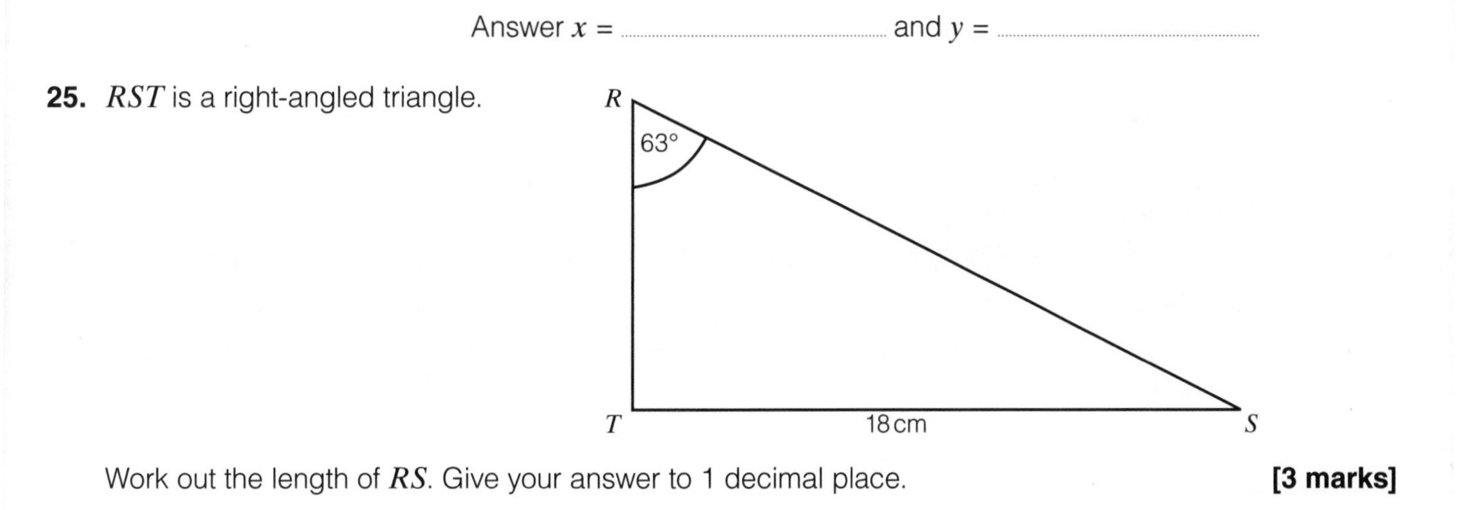

Work out the length of RS. Give your answer to 1 decimal place. **[3 marks]**

Answer .... cm

END OF QUESTIONS

Letts

GCSE

Mathematics

F

Foundation tier

Paper 3

Time: 1 hour 30 minutes

For this paper you must have:

- a calculator
- mathematical instruments

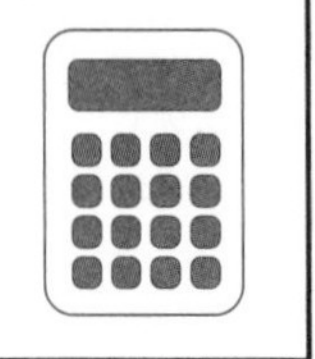

Instructions

- Use black ink or black ball-point pen. Draw diagrams in pencil.
- Read each question carefully before you start to write your answer.
- Diagrams are **not** accurately drawn unless otherwise stated.
- Answer **all** the questions.
- Answer the questions in the space provided.
- In all calculations, show clearly how you work out your answer. Use a separate sheet of paper if needed. Marks may be given for a correct method even if the answer is wrong.
- If your calculator does not have a π button, take the value of π to be 3.142 unless the question instructs otherwise.

Information

- The marks for each question are shown in brackets.
- The maximum mark for this paper is 80.

Name: ....

1. **(a)** Write the number three thousand seven hundred in figures. **[1 mark]**

Answer ....

(b) Write down the value of 6 in the number 5 432 679. **[1 mark]**

Answer ....

(c) Which digit is in the tenths position in the number 23.457? **[1 mark]**

Answer ....

2. **(a)** Write down a factor of 12 which is also a prime number. **[2 marks]**

Answer ....

(b) Write down an odd square number which is greater than 30. **[1 mark]**

Answer ....

3. A, B and C are three vertices of a rectangle.

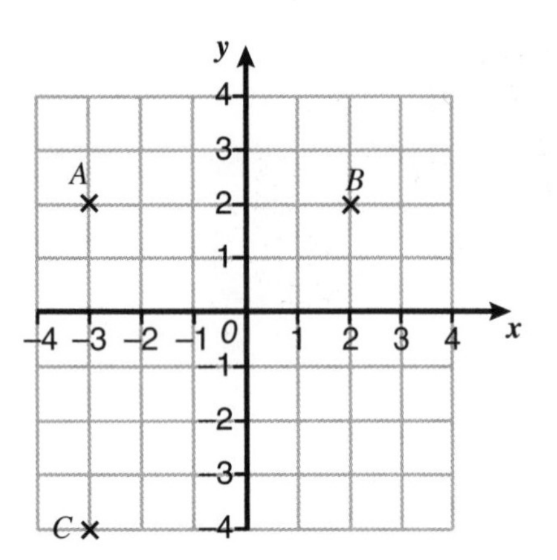

(a) Write down the coordinates of the missing vertex. **[1 mark]**

Answer (.........., ...)

(b) Write down the coordinates of the midpoint of the line AB. **[2 marks]**

Answer (.........., ...)

4. Expand $4a(2a - b)$ **[2 marks]**

Answer ....

5. Ro weighs himself at a keep-fit club. The arrow on the scale shows his weight.

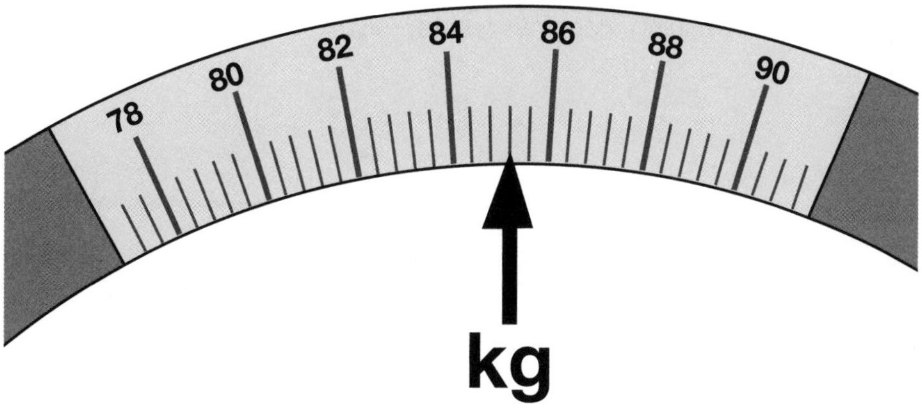

78 80 82 84 86 88 90

kg

He has lost 1.4 kg this week.

Last week he lost 0.8 kg

How much did Ro weigh two weeks ago? **[3 marks]**

Answer .... kg

6. There are 25 students in Tilac's class.

- 9 students are boys.
- 15 students do not wear glasses.

(a) What fraction of Tilac's class wear glasses? Write your answer in its simplest form. **[2 marks]**

Answer ....

(b) What percentage of Tilac's class are girls? **[2 marks]**

Answer .... %

7. Twelve identical washing machines weigh a total of 1625.22 kg.

(a) How much do 15 of the same washing machines weigh? Give your answer correct to 3 decimal places. **[2 marks]**

Answer .... kg

(b) A lorry can carry a total load of 2.6 tonnes.

What is the maximum number of these washing machines it can carry? **[2 marks]**

Answer ....

8. The following calculation has missing symbols.

3 [?] 4 [?] 5 [?] 6 = 4.5

Use any of the symbols +, −, ×, ÷ to make the calculation correct.

You may also use brackets ( ) if needed. **[2 marks]**

Answer ....

9. The total cost (C) to learn to drive is:

The number of lessons taken (n) multiplied by £25 Plus The number of tests taken (t) multiplied by £85

(a) Write a formula for C in terms of n and t. **[2 marks]**

Answer ....

(b) Lorna has 17 lessons and passes the test on her second attempt.

Find the total cost for Lorna to learn to drive. **[2 marks]**

Answer £ ....

(c) Vic passed his test on the first attempt. It cost him £560 to learn to drive.

How many lessons did Vic have? **[2 marks]**

Answer ....

10. 1 inch = 2.54 cm

Jacob measures the length of his room as 655 cm.

Find the length of Jacob's room to the nearest inch. **[2 marks]**

Answer .... inches

11. Lauren sells camping goods online.

The probability that she has an item returned is 6.4%

Next month Lauren expects to sell 250 items.

How many items should she expect to be returned? **[2 marks]**

Answer ....

12. Kayley is 4 years older than Micah.

Micah is double the age of Rebekah.

Rebekah is x years old.

(a) Write an expression, in terms of x, for Kayley's age. **[2 marks]**

Answer .... years old

The total age of Kayley, Micah and Rebekah is 19 years.

(b) How old is Micah? **[3 marks]**

Answer .... years old

13. Carl can run 100 m in 10 seconds.

Work out his speed in kilometres per hour. **[3 marks]**

Answer .... km/h

14. This L-shaped garden is going to be covered in turf and made into a lawn.

Turf is sold at £3.99 per m².

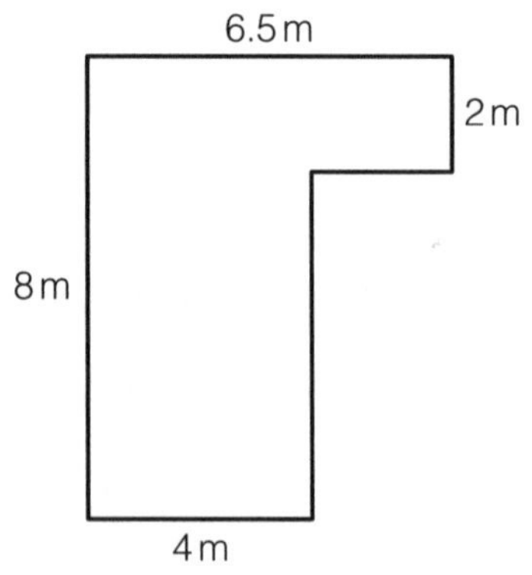

6.5 m
2 m
8 m
4 m

Work out the cost of covering the garden with turf. **[4 marks]**

Answer £ ....

15. John invests £5450 at 3.5% per annum for four years.

Work out the simple interest that he will receive. **[2 marks]**

Answer £ ....

16. This right-angled triangle has an area of 6 cm^2 and a perimeter of 12 cm.

.... cm

.... cm

.... cm

On the diagram, write down the length of the three sides, given that each side is an integer. **[3 marks]**

17. **(a)** Write 1260 as a product of its prime factors. Give your answer in index form. **[3 marks]**

Answer ....

(b) Find the highest common factor (HCF) of 1260 and 1050. **[2 marks]**

Answer ....

18. Carolyn, Del and Julien are playing computer games.

Carolyn has four times as many points as Del.

Del has $\frac{1}{3}$ of the points that Julien has.

Write down the ratio of points that Carolyn, Del and Julien have. **[2 marks]**

Answer ....

19. Tim and Nat have a race. Tim cycles while Nat runs.

Nat gets a 20-second head start.

They both finish the race at exactly the same time.

The velocity–time graph shows the start of the race and it continues beyond the end of the race.

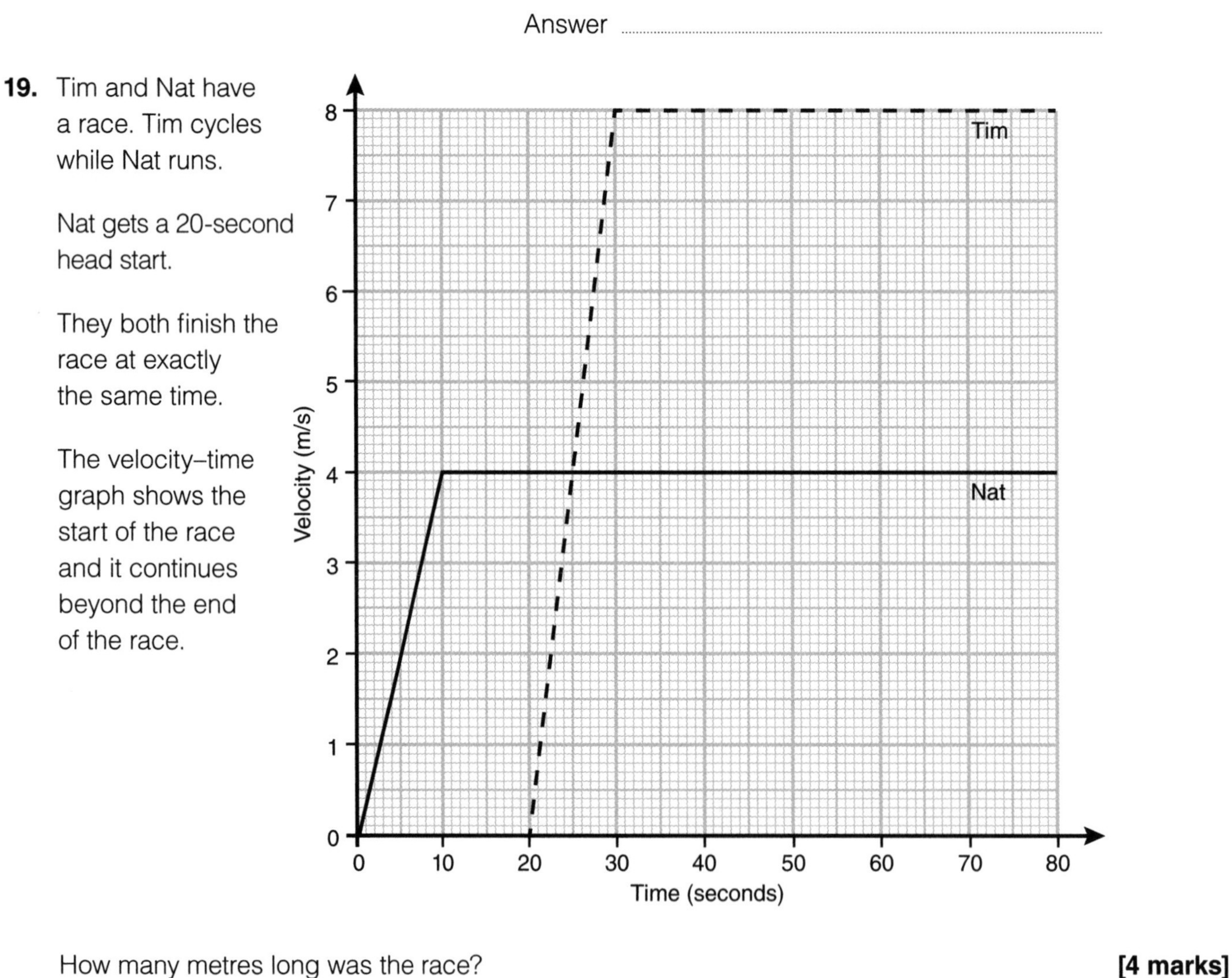

How many metres long was the race? **[4 marks]**

Answer .... m

20. A submarine (S) is marked on the scale diagram. Rocks (R) are also marked on the diagram.

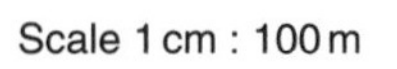

R

N

S

(a) Find the bearing of the rocks from the submarine. **[2 marks]**

Answer .... °

(b) A sunken galleon is on a bearing of 076° from the submarine at a distance of 500 m.

Mark the position of the galleon (G) on the diagram. **[2 marks]**

(c) The submarine is going to pass between the rocks and the galleon. It must pass:

- Closer to the galleon than the rocks.
- At least 350 m away from the galleon.

Shade the region through which the submarine can pass. **[2 marks]**

21. Jill and David are both artists. Here is some information about the price and the number of paintings they sold over a 12-month period.

Jill

Price, £	**Frequency**
$0 < £ \leqslant 100$	3
$100 < £ \leqslant 200$	8
$200 < £ \leqslant 300$	6
$300 < £ \leqslant 400$	7

David

Price, £	**Frequency**
$0 < £ \leqslant 100$	2
$100 < £ \leqslant 200$	4
$200 < £ \leqslant 300$	8
$300 < £ \leqslant 400$	6

On average, who earned the most money per painting? Show your working. **[5 marks]**

Answer ....

22. Here are six graphs. Each one is labelled.

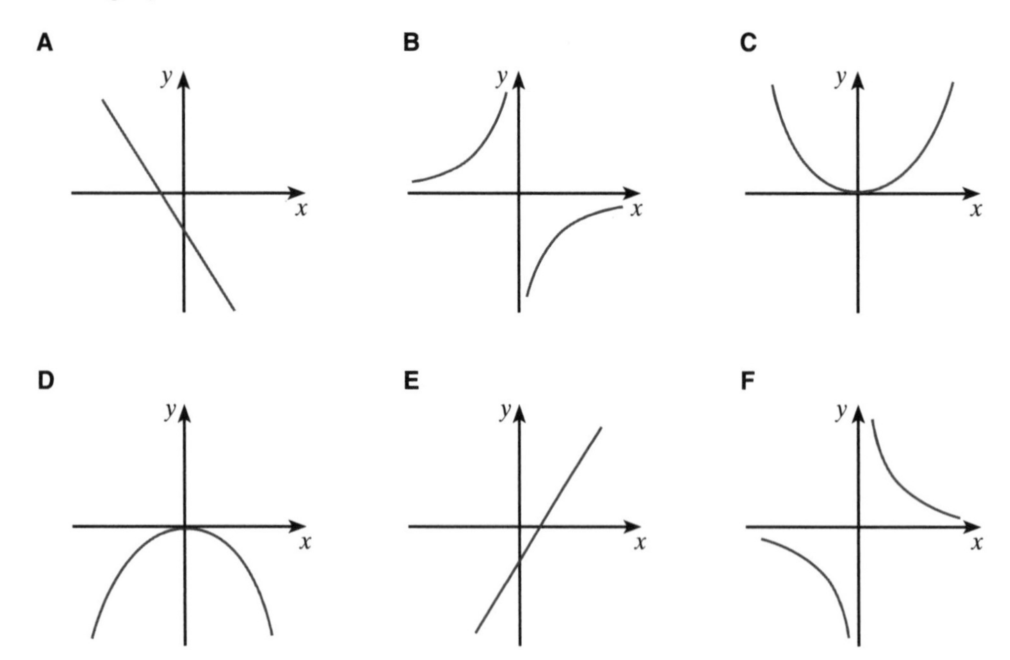

Write down the letter of the graph which matches these equations.

(a) $y = 3x - 2$ Answer .... **[1 mark]**

(b) $y = x^2$ Answer .... **[1 mark]**

(c) $y = \frac{1}{x}$ Answer .... **[1 mark]**

23. Jason is waiting for a delivery of concrete and a delivery of roof slates for his building project.

The probability that his concrete will arrive on time is 0.75

The probability that his roof slates will arrive on time is 0.65

(a) Complete the probability tree diagram. **[2 marks]**

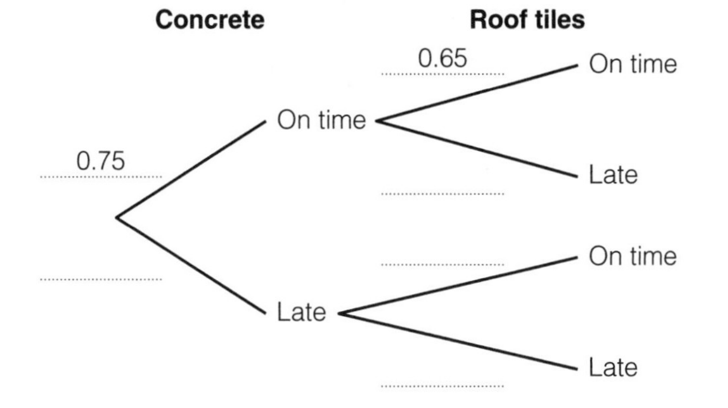

(b) Find the probability that both of the deliveries will arrive late. **[2 marks]**

Answer ....

END OF QUESTIONS

END OF PRACTICE EXAM 1

Letts

GCSE

Mathematics

F

Foundation tier

Paper 1

Time: 1 hour 30 minutes

For this paper you must have:

- mathematical instruments

You must **not** use a calculator.

Instructions

- Use black ink or black ball-point pen. Draw diagrams in pencil.
- Read each question carefully before you start to write your answer.
- Diagrams are **not** accurately drawn unless otherwise stated.
- Answer **all** questions.
- You must answer the questions in the space provided.
- In all calculations, show clearly how you work out your answer. Use a separate sheet of paper if needed. Marks may be given for a correct method even if the answer is wrong.

Information

- The marks for each question are shown in brackets.
- The maximum mark for this paper is 80.

Name: ....

1. The range of the following six numbers is 5.

5	9	8	5	7	?

(a) Write down a possible value for the missing number. **[1 mark]**

Answer ....

(b) Write down a **different** possible value for the missing number. **[1 mark]**

Answer ....

(c) Write down the mode for the six numbers. **[1 mark]**

Answer ....

2. **(a)** Which **one** of the following numbers is prime?

Circle your answer. **[1 mark]**

1 6 9 13 15

(b) Which **one** of the following is a cube number?

Circle your answer. **[1 mark]**

2 4 6 8 10

3. **(a)** Simplify $3a + 2b - a - 3b$ **[2 marks]**

Answer ....

(b) Simplify $4x \times 5y$

Circle your answer. **[1 mark]**

$9x^2y^2$ $20xy$ $9xy$ $20xy^2$

4. Find 21% of 200. **[2 marks]**

Answer ....

5. Nathan has a one-hour lunch break. He spends 35 minutes in the gym.

What fraction of his lunch break is spent in the gym? Write your answer in its simplest form. **[2 marks]**

Answer ....

6. Place the following in order from smallest to largest.

$\frac{12}{36}$ 32.5% $\frac{8}{25}$ 0.43 **[2 marks]**

Answer ....

7. Ashley's deluxe hot chocolate costs £1.65

He pays the exact amount with six coins, none of which are 1p or 2p coins.

Give two possibilities for the six coins he uses. **[2 marks]**

Possibility 1: ....

Possibility 2: ....

8. Rectangle A is cut along its diagonal to make triangle B.

A

B

Place one of the symbols $>$, $<$ or $=$ in the box to make each statement correct.

(a) The area of B is ☐ half of the area of A. **[1 mark]**

(b) The perimeter of B is ☐ half of the perimeter of A. **[1 mark]**

9. A jeweller has 0.35 kg of gold. He is making gold rings, each of which contain 12 g of gold.

(a) What is the maximum number of gold rings he can make? **[3 marks]**

Answer ....

(b) How much gold will be left over? **[1 mark]**

Answer .... g

10. Is the following calculation correct?

$$\frac{5}{6} - \frac{1}{3} = \frac{4}{3}$$

Give reasons for your answer. **[2 marks]**

....

....

....

11. Look at the following expressions:

$$\frac{a-b}{a} \qquad \frac{ab}{a} \qquad \frac{b-a}{a}$$

When $a = -2$ and $b = -3$, which expression gives a positive value? **[3 marks]**

Answer ....

12. Is 45 a term in the sequence $5n - 3$? Give reasons for your answer. **[2 marks]**

....

....

....

....

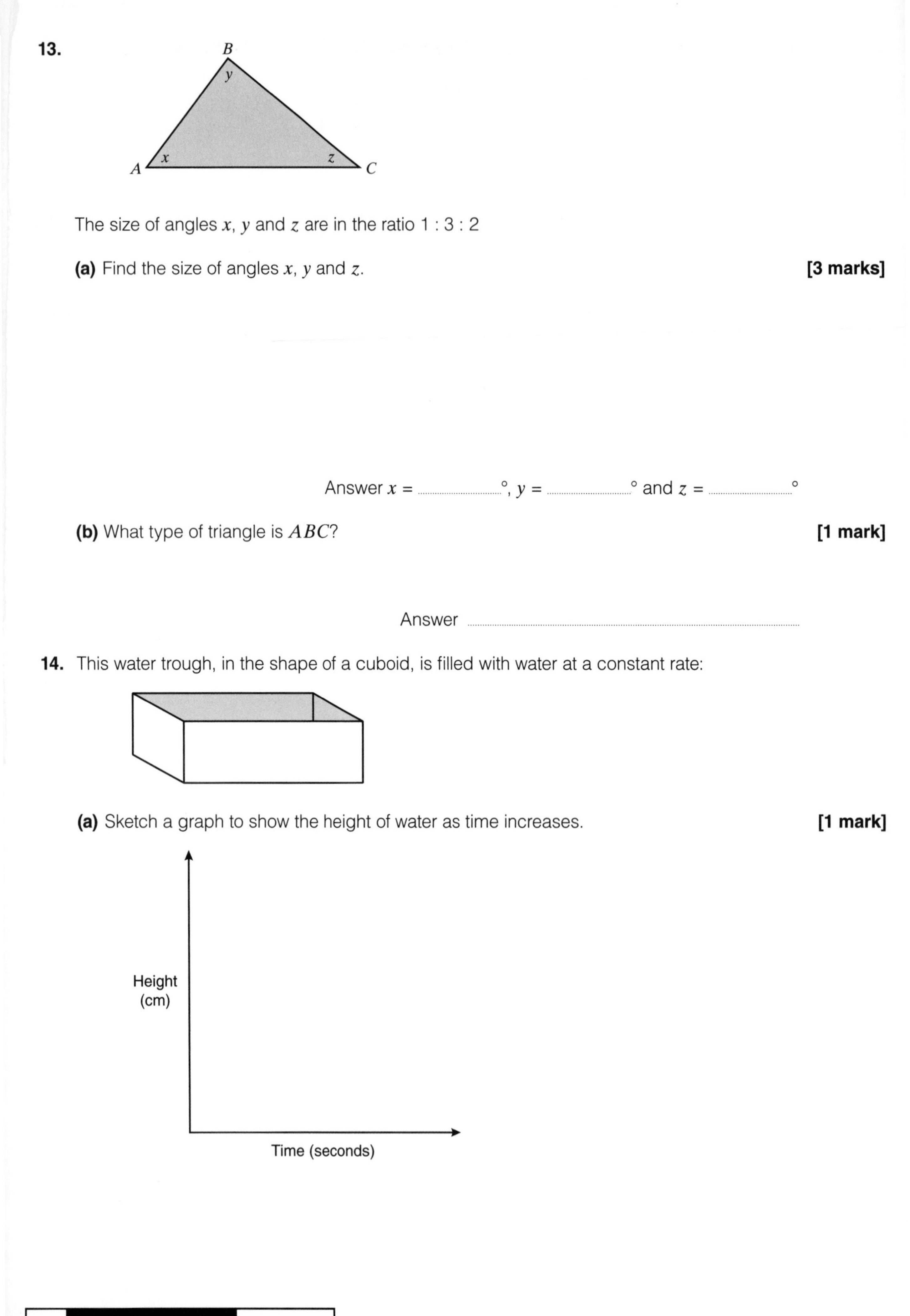

13.

The size of angles x, y and z are in the ratio 1 : 3 : 2

(a) Find the size of angles x, y and z. **[3 marks]**

Answer x = ...°, y = ...° and z = ...°

(b) What type of triangle is ABC? **[1 mark]**

Answer ....

14. This water trough, in the shape of a cuboid, is filled with water at a constant rate:

(a) Sketch a graph to show the height of water as time increases. **[1 mark]**

(b) A different water trough, in the shape of half a cylinder, is filled with water at a constant rate:

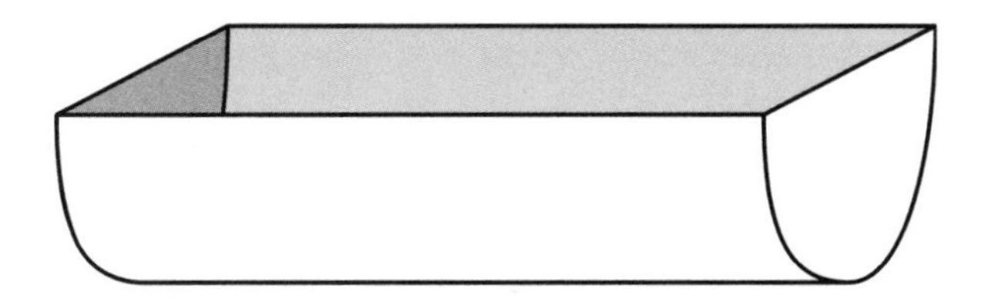

Sketch a graph to show the height of water as time increases. **[1 mark]**

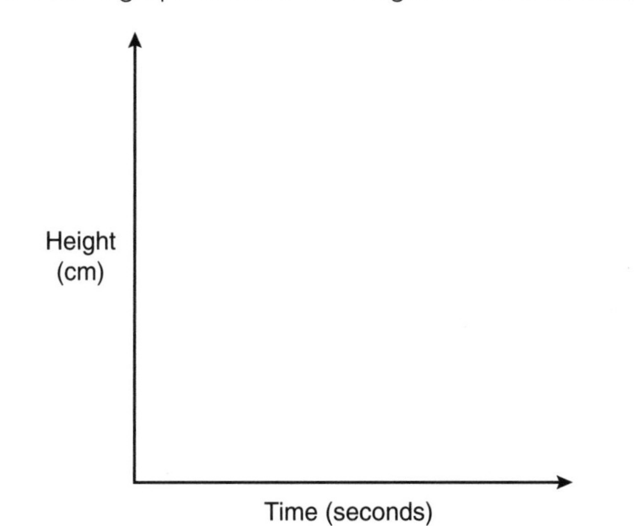

15. The line AB represents a distance of 20 km.

A ———————————————————— B

What distance does 1 cm represent in the diagram? **[2 marks]**

Answer .... km

16. A straight line graph is shown.

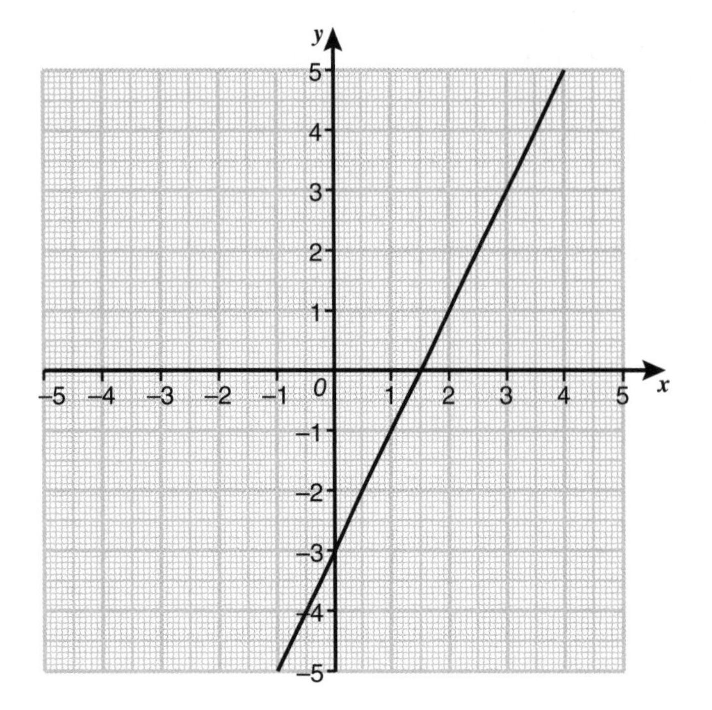

Find the equation of the line in the form $y = mx + c$ **[3 marks]**

Answer ....

17.

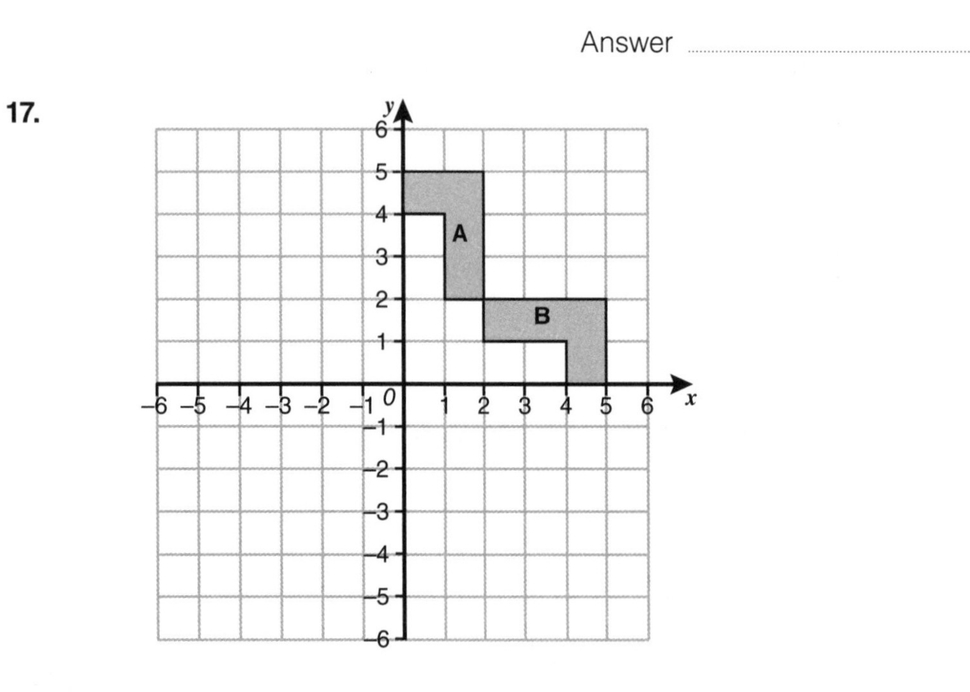

(a) Describe fully the transformation which maps shape A onto shape B. **[2 marks]**

....

....

(b) Rotate shape A 90° anticlockwise about the point (–1, 1). **[2 marks]**

18. A football team plays 46 matches in a season. 21 are at home.

The team wins 15 of its home games and loses 2.

The team loses 12 of its away games and wins 10.

(a) Complete the frequency tree diagram to represent this data. **[2 marks]**

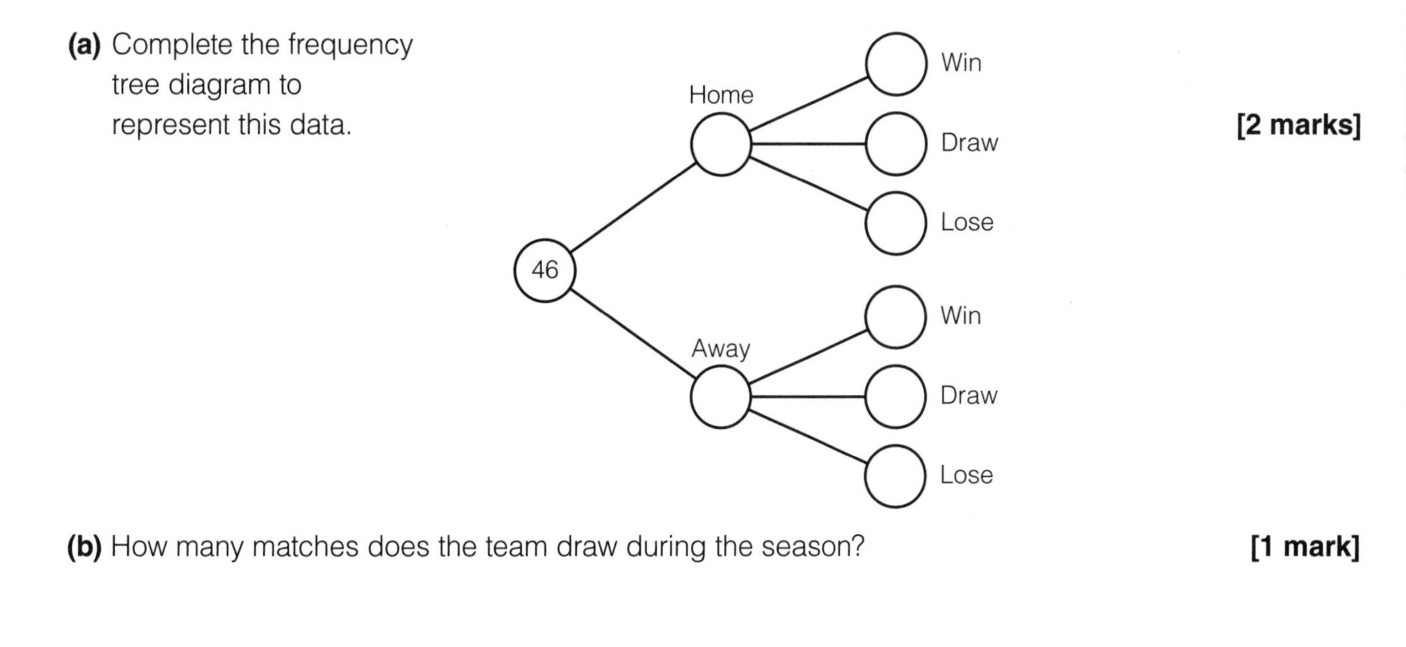

(b) How many matches does the team draw during the season? **[1 mark]**

Answer ....

19. This table shows the average diameter of blood vessels in the human body.

Blood vessel	Length (mm)
Vein	2.44×10^{-2}
Artery	5.25×10^{-2}
Capillary	3.12×10^{-3}
Small artery	1.82×10^{-2}
Venule	5.19×10^{-3}
Small vein	1.43×10^{-2}
Arteriole	7.8×10^{-3}

(a) Write down the average diameter of the venule as an ordinary number. **[1 mark]**

Answer .... mm

(b) Write down the name of the blood vessel which has the smallest diameter and the name of the blood vessel which has the largest diameter. **[2 marks]**

Smallest: ....

Largest: ....

20. Fifty people in a bike shop are asked what type of bike they own.

The Venn diagram shows the results of the survey.

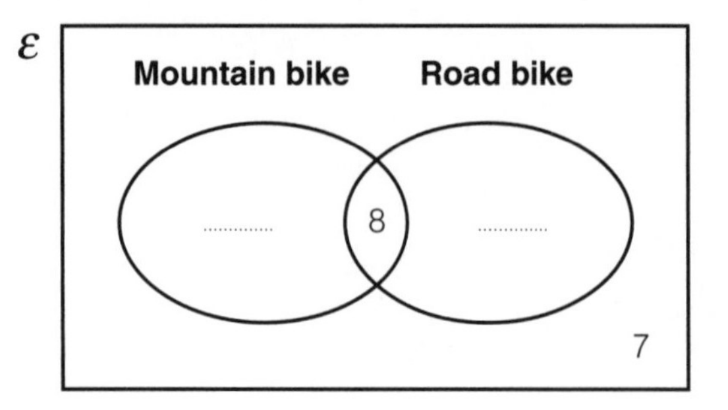

Of the 50 people questioned, 22 people own a mountain bike.

(a) Complete the Venn diagram. **[2 marks]**

(b) Someone from the survey is chosen at random.

What is the probability that the person owns a road bike? **[1 mark]**

Answer ....

21. Work out 27×2.36 **[3 marks]**

Answer ....

22. Solve $5(x - 4) = 3x - 15$ **[3 marks]**

Answer $x =$ ....

23. Find $\frac{3}{5}$ of $3\frac{3}{4}$ **[3 marks]**

Answer ....

24. Julie set up a video camera in her garden. Over 30 nights she recorded how many different times she captured some wild animals on camera.

Here are her results:

Animal	**Frequency (number of nights seen)**
Fox	8
Badger	2
Owl	3
Stoat	5
Hedgehog	11
Other rodent	12

(a) On any given night, find the probability that Julie will record a fox. **[1 mark]**

Answer ....

(b) On any given night, find the probability that she will record an owl and a stoat. **[2 marks]**

Answer ....

25. There are 350 trees in a forest. After strong winds, only 280 trees are left standing.

What percentage of trees has been destroyed? **[2 marks]**

Answer ....%

26. Jonathon is driving from Manchester to Stoke-on-Trent. On the way, he stops off at Knutsford services. The total distance for the journey from Manchester to Stoke-on-Trent is 66 miles.

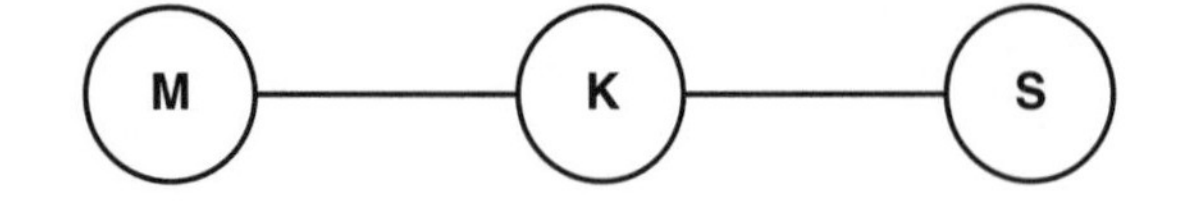

The journey from Manchester to Knutsford services takes 40 minutes and Jonathon travels at an average speed of 51 mph. The journey from Knutsford services to Stoke-on-Trent takes 30 minutes.

Find his average speed for the second part of the journey. **[3 marks]**

Answer .... mph

27. Here are four triangles: ABC, DEF, GHI and JKL.

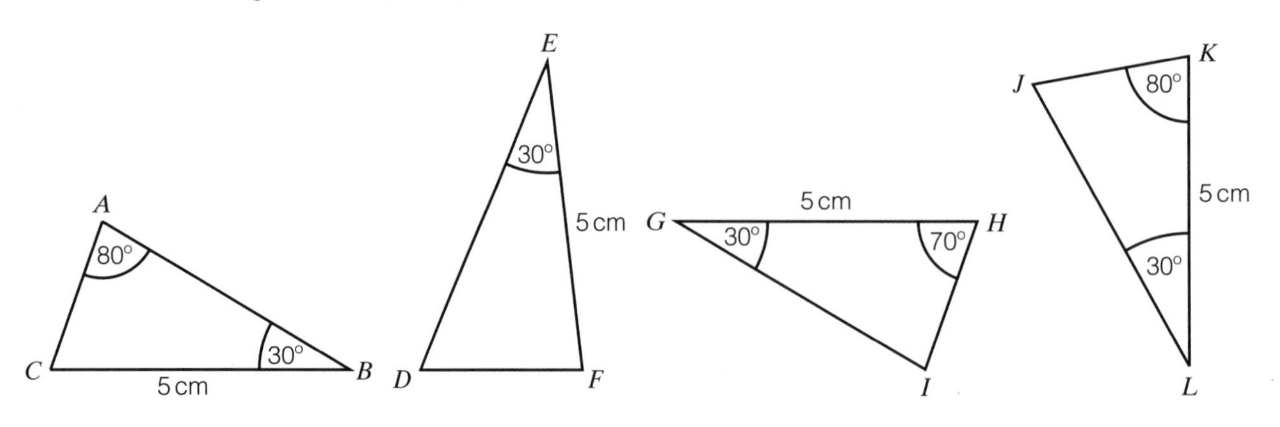

Which two triangles are congruent? Give reasons. **[2 marks]**

....

....

....

28. This quadrant has radius $2x$.

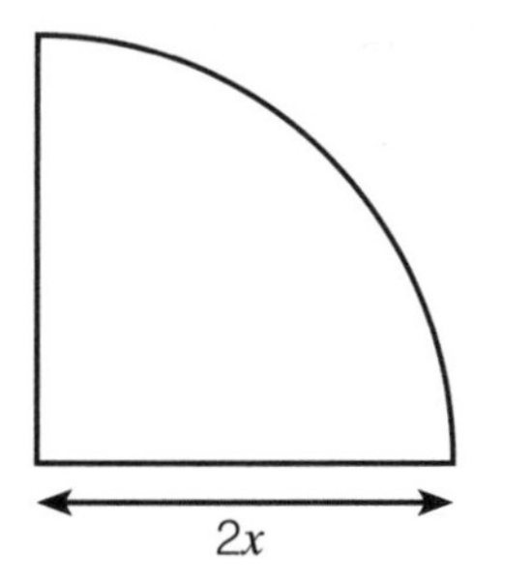

Find the perimeter of the quadrant. Give your answer in terms of x and π. **[2 marks]**

Answer ....

29. $a = \begin{pmatrix} -2 \\ q \end{pmatrix}$ $b = \begin{pmatrix} r \\ 8 \end{pmatrix}$ $a + 2b = \begin{pmatrix} 4 \\ 21 \end{pmatrix}$

Write down the value of q and r. **[2 marks]**

Answer $q =$ .... and $r =$ ....

30. Solve $x^2 - 9x - 36 = 0$ **[2 marks]**

Answer $x =$ .... and $x =$ ....

31. Kevin walks from Land's End to John O'Groats. He calculates that the distance will be 880 miles to the nearest 5 miles.

Write down the error interval for the distance in miles, m. **[2 marks]**

Answer ....

END OF QUESTIONS

Letts

GCSE

Mathematics

Foundation tier

F

Paper 2

Time: 1 hour 30 minutes

For this paper you must have:

- a calculator
- mathematical instruments

Instructions

- Use black ink or black ball-point pen. Draw diagrams in pencil.
- Read each question carefully before you start to write your answer.
- Diagrams are **not** accurately drawn unless otherwise stated.
- Answer **all** the questions.
- Answer the questions in the space provided.
- In all calculations, show clearly how you work out your answer. Use a separate sheet of paper if needed. Marks may be given for a correct method even if the answer is wrong.
- If your calculator does not have a π button, take the value of π to be 3.142 unless the question instructs otherwise.

Information

- The marks for each question are shown in brackets.
- The maximum mark for this paper is 80.

Name: ....

1. **(a)** Round 34 562 to the nearest 100. **[1 mark]**

Answer ....

(b) Round 428 098 to the nearest 1000. **[1 mark]**

Answer ....

2. Write the following decimals in order of size from smallest to largest.

0.803 0.083 0.83 0.8 0.883 **[2 marks]**

Answer ....

3. The table and bar chart show the colour of cars in a car park at midday.

Colour of car	Red	White	Silver	Blue	Green	Black	Maroon	Other
Frequency	8		10		4		2	3

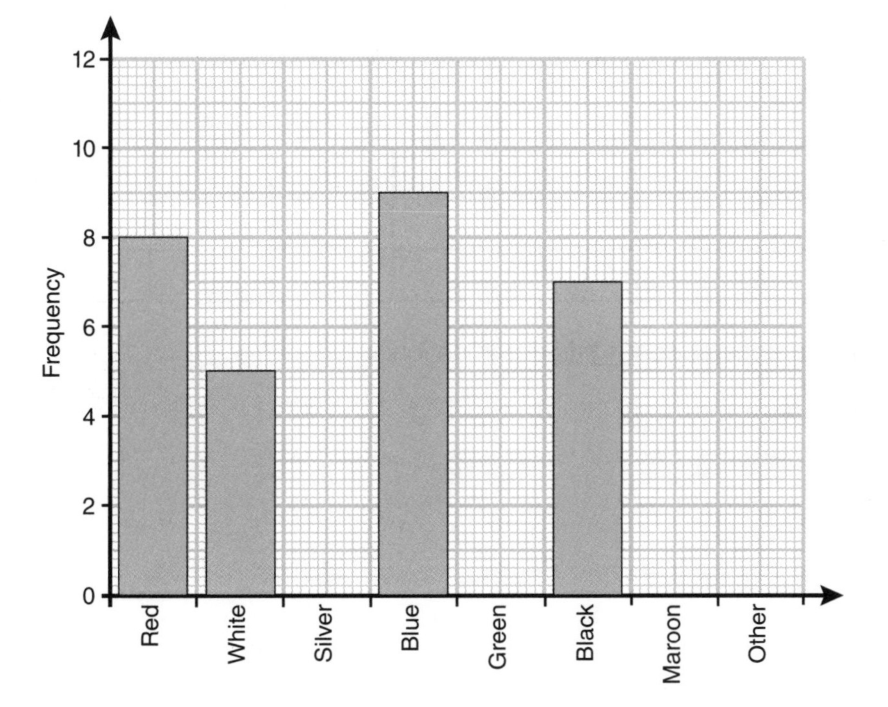

(a) Complete the frequency table and bar chart. **[2 marks]**

An hour later, 5 cars have left and 8 have arrived.

(b) How many cars are now in the car park? **[1 mark]**

Answer ....

4. **(a)** Calculate 4^4

Circle your answer. **[1 mark]**

8   16   64   256

(b) Calculate $\sqrt{529}$ **[1 mark]**

Answer ....

(c) Calculate $\sqrt[3]{125}$

Circle your answer. **[1 mark]**

5   25   75   41.666...

5. A mobile phone company offers different options for minutes, texts and data when choosing a new contract. The table shows the options available.

Option	Minutes	Texts	Data
A	150	5000	500 MB
B	500	Unlimited	1 GB

When choosing a new contract, you can choose any combination of minutes, texts and data.

(a) Write down all of the different possible combinations that someone could choose.
The first two are done for you. **[2 marks]**

Minutes	A	A						
Texts	A	A						
Data	A	B						

(b) Find the probability that someone chooses this combination:

500 minutes, 5000 texts and 1 GB of data **[1 mark]**

Answer ....

6. Lee receives the following phone bill for October.

Total call time	Cost
3 hours 45 minutes	£41.40

Find the average cost of his calls per minute. **[2 marks]**

Answer ....

7. Look at this shape:

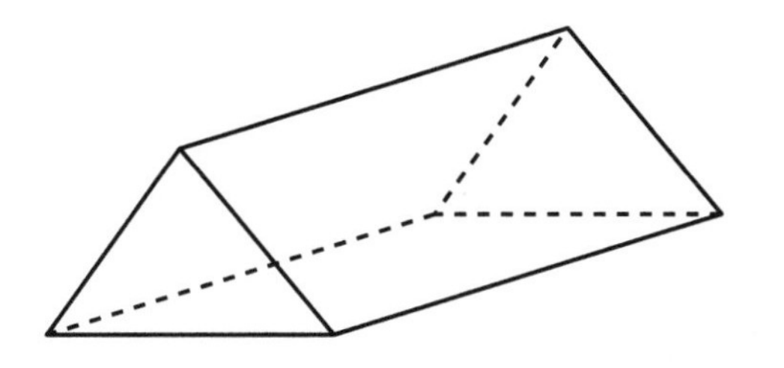

(a) Name the three-dimensional shape shown. **[1 mark]**

Answer ....

(b) Write down the number of faces, edges and vertices that the shape has. **[2 marks]**

Faces: ....

Edges: ....

Vertices: ....

8. The graph enables you to convert between degrees Celsius and degrees Fahrenheit.

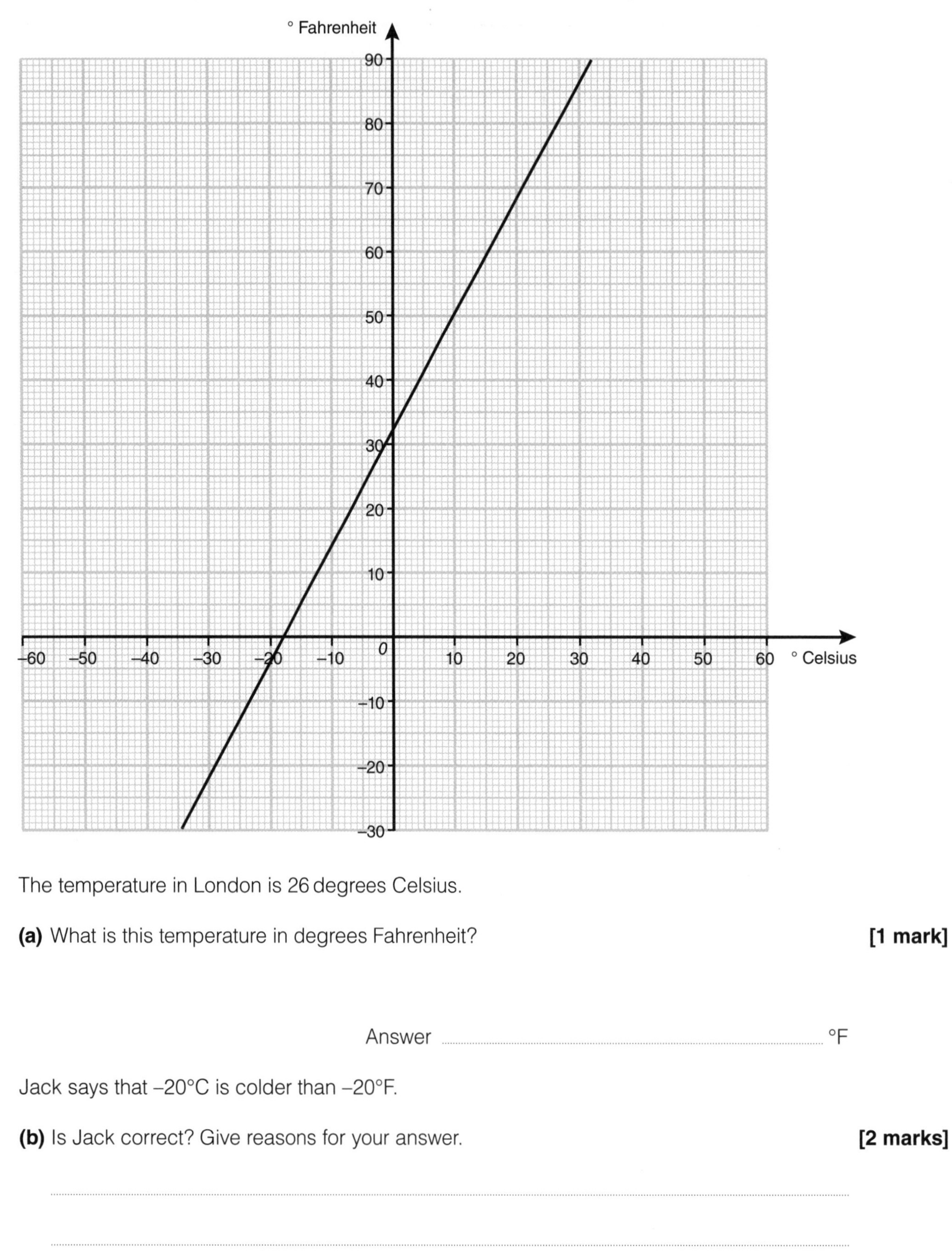

The temperature in London is 26 degrees Celsius.

(a) What is this temperature in degrees Fahrenheit? **[1 mark]**

Answer .... °F

Jack says that –20°C is colder than –20°F.

(b) Is Jack correct? Give reasons for your answer. **[2 marks]**

....

....

9.

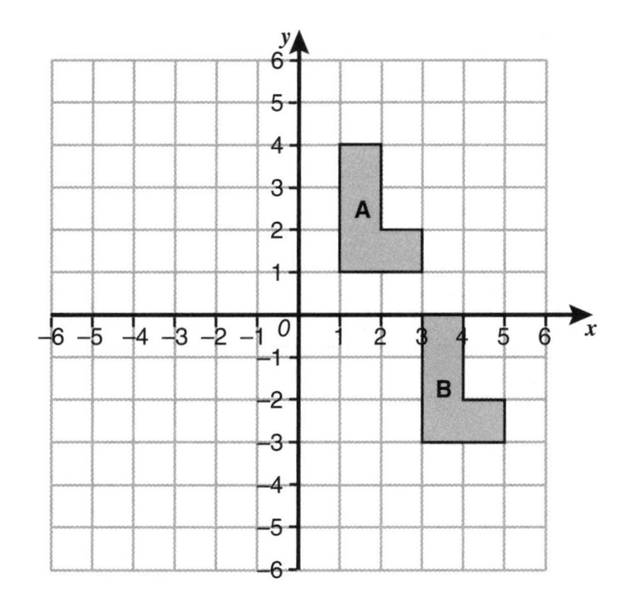

Describe the single transformation which maps shape A onto shape B. **[2 marks]**

....

....

10. In a game, a prize is won if you score a number 6. You can choose from:

Option 1: Rolling a fair, six-sided dice

or

Option 2: Spinning this spinner

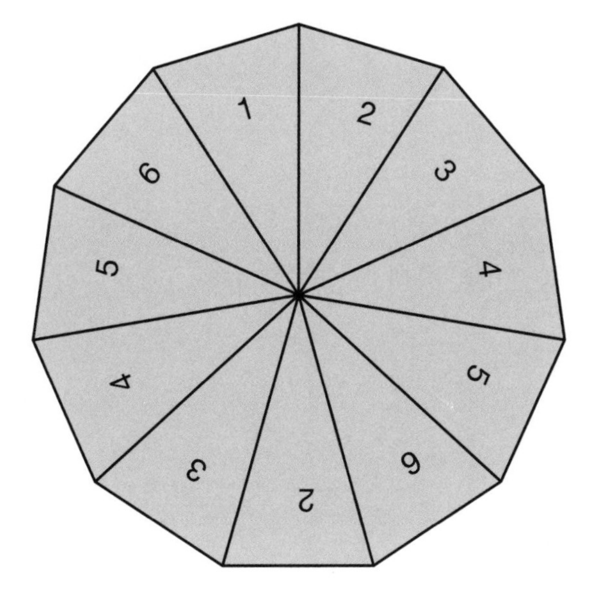

Which option gives the best chance of winning a prize? Give a reason for your answer. **[3 marks]**

Answer ....

Reason ....

....

11. Lyndsey wants to buy three bottles of the same shampoo.

The cost of the bottle at Booties is £3.59

The cost of the bottle at Hair-Care is £3.79

The two shops offer different promotions on the shampoo:

Booties
Buy two, get one free

Hair-Care
40% off your total bill

Which shop gives the best value for money? You must show your working. **[4 marks]**

Answer ....

12. These are the ingredients needed to make 25 biscuits:

- 150 g of self-raising flour
- 150 g of caster sugar
- 120 g of rolled oats
- 150 g of butter
- 15 g of syrup

Work out how much of each ingredient is needed to make 40 of these biscuits. **[3 marks]**

Self-raising flour: .... g

Caster sugar: .... g

Rolled oats: .... g

Butter: .... g

Syrup: .... g

13. The lengths of two different types of British snakes were recorded. These are the results:

Adder

Length, l (cm)	Frequency
$50 < l \leqslant 55$	4
$55 < l \leqslant 60$	9
$60 < l \leqslant 65$	11
$65 < l \leqslant 70$	8
$70 < l \leqslant 75$	3

Grass snake

Length, l (cm)	Frequency
$50 < l \leqslant 55$	3
$55 < l \leqslant 60$	5
$60 < l \leqslant 65$	11
$65 < l \leqslant 70$	13
$70 < l \leqslant 75$	7

Is the median length for the adder and the grass snake in the same class interval?
Show working to support your conclusions. **[3 marks]**

Answer ....

14. Jason teaches piano lessons. He has 30 students.

He charges £16 per lesson for children and £21 a lesson for adults.

$\frac{4}{5}$ of his students are children.

He teaches $\frac{1}{6}$ of the children twice a week. He teaches all of the adults once a week.

How much money does Jason earn in a week? **[5 marks]**

Answer £ ....

15. Stu wants to find out how many fish there are in a small lake.

One day he catches 40 fish. He marks them and puts them back.

The next day he catches 60 fish. Two of them are marked.

Estimate how many fish may be in the lake. **[3 marks]**

Answer ....

16. (a) Solve the equation $\frac{2x + 3}{4} = 2.5$ **[2 marks]**

Answer $x =$ ....

(b) Make u the subject of $v^2 = u^2 + 2as$ **[2 marks]**

Answer ....

17. Last year, Instatwitt had 1 372 658 users. This year it has 1 587 392 users.

What percentage increase is this? Give your answer to 1 decimal place. **[2 marks]**

Answer .... %

18. Jessica and Peter share shoelace sweets in the ratio of 11 : 15

Peter gets two more sweets than Jessica.

How many sweets does Peter get? **[3 marks]**

Answer .... sweets

19. 162 students from Year 10 are having a reward day.

66.$\dot{6}$% go to a theme park	$\frac{8}{27}$ go ice-skating	The rest stay in school

How many students are staying in school? **[4 marks]**

Answer .... students

20. **(a)** Write the first five terms of the following term-to-term sequence.

'Start with the first prime number, then add five each time.' **[1 mark]**

Answer ....

(b) Find the nth term rule for the sequence. **[2 marks]**

Answer ....

(c) Find the 27th term in the sequence. **[1 mark]**

Answer ....

21. A wedding venue uses eight chefs to prepare a meal for 140 guests.

(a) How many chefs would be needed to prepare a meal for 175 guests? **[2 marks]**

Answer .... chefs

(b) Write down one assumption which you have made. **[1 mark]**

....

....

22. An ice-cube tray holds 16 ice cubes with side length 2 cm.

Shiv tips all 16 ice cubes into a cylindrical jug with diameter 12 cm.

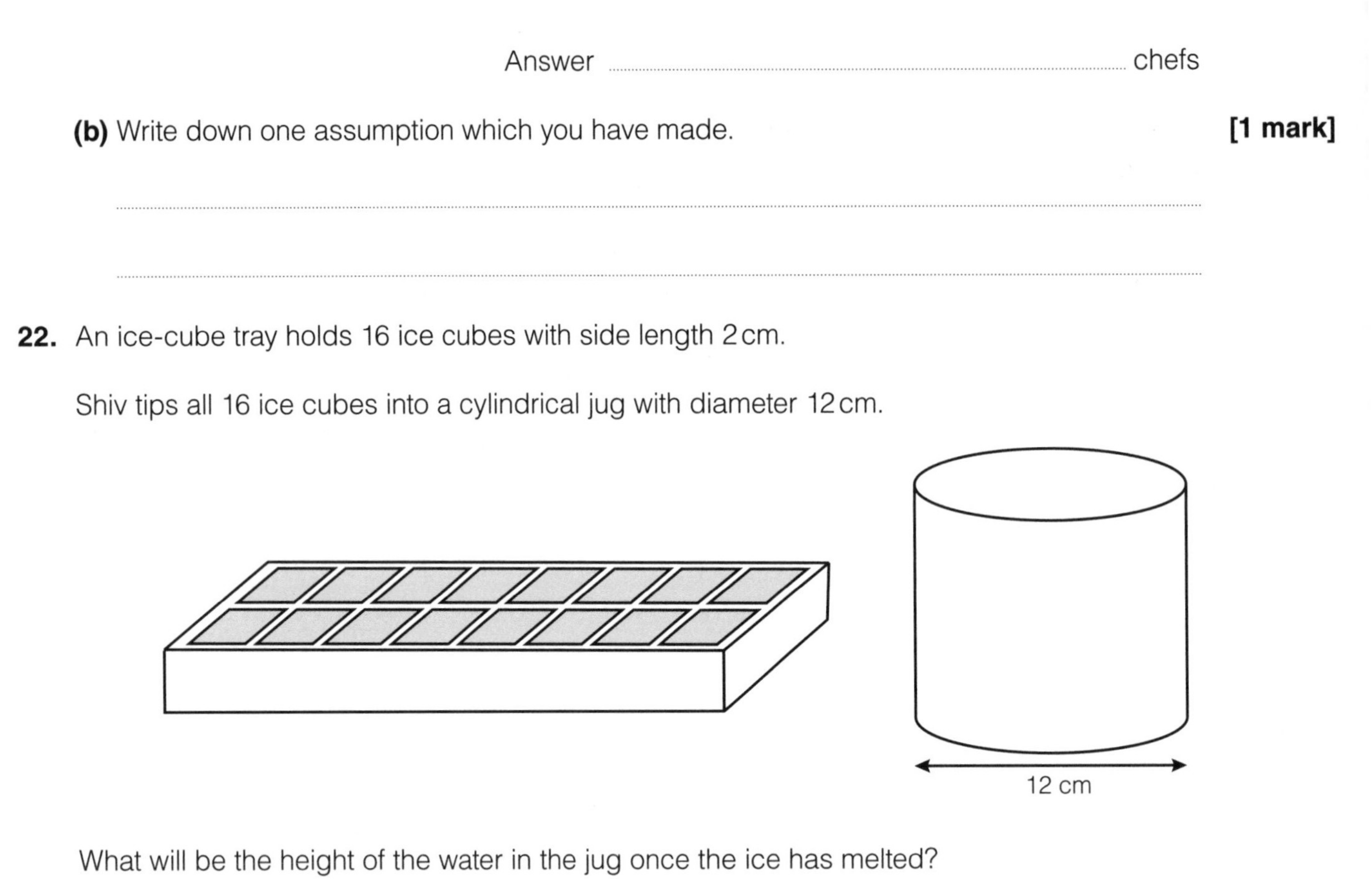

What will be the height of the water in the jug once the ice has melted?
Give your answer to 3 significant figures. **[4 marks]**

Answer .... cm

23. **(a)** Sketch the graph of $y = x^2 - 2$ **[2 marks]**

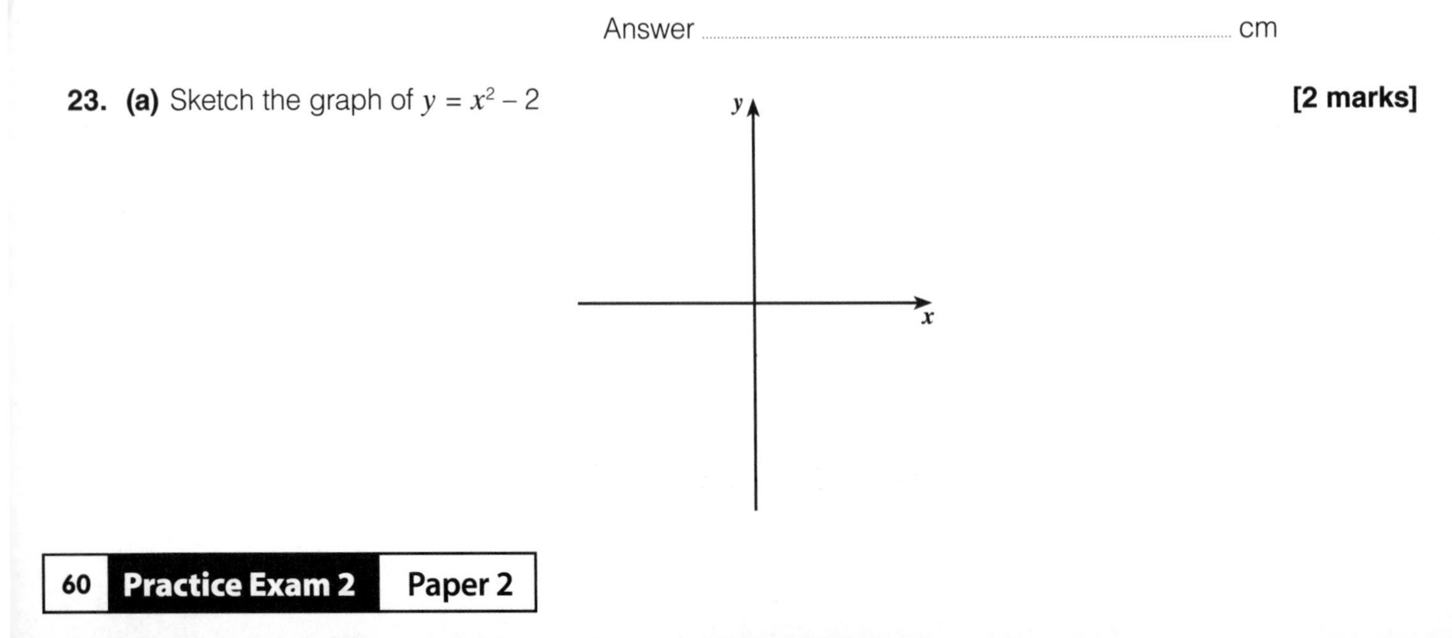

(b) A straight line passes through the two points (3, 2) and (5, 6).

Find the equation of the line. **[3 marks]**

Answer ....

24. A man is standing at the top of some vertical cliffs looking out to sea.

His measuring instrument can measure the angle of depression and the distance from himself to a boat.

The angle of depression is 62°. The distance is 658 m.

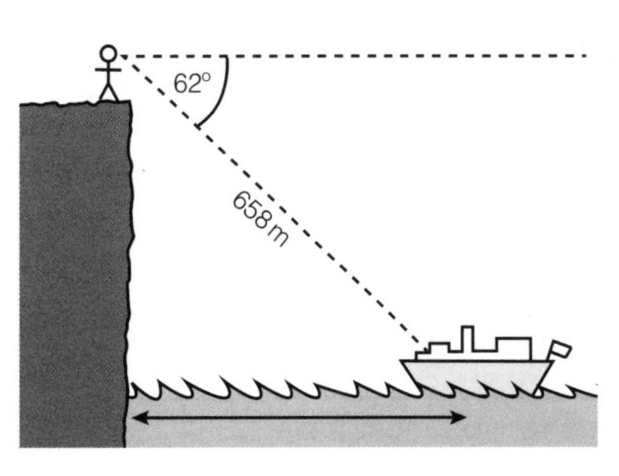

Find the horizontal distance from the boat to the cliffs. Give your answer correct to 3 significant figures. **[3 marks]**

Answer .... m

25. **(a)** Evaluate $\frac{6^7 \times 6^{-2}}{6^6}$ **[2 marks]**

Answer ....

(b) Find the value of x when $8^x = 16$ **[2 marks]**

Answer $x =$ ....

END OF QUESTIONS

GCSE

Mathematics

F

Foundation tier

Paper 3

Time: 1 hour 30 minutes

For this paper you must have:

- a calculator
- mathematical instruments

Instructions

- Use black ink or black ball-point pen. Draw diagrams in pencil.
- Read each question carefully before you start to write your answer.
- Diagrams are **not** accurately drawn unless otherwise stated.
- Answer **all** the questions.
- Answer the questions in the space provided.
- In all calculations, show clearly how you work out your answer. Use a separate sheet of paper if needed. Marks may be given for a correct method even if the answer is wrong.
- If your calculator does not have a π button, take the value of π to be 3.142 unless the question instructs otherwise.

Information

- The marks for each question are shown in brackets.
- The maximum mark for this paper is 80.

Name: ....

1. (a) Solve $x - 3 = 9$ **[1 mark]**

Answer $x =$ ....

(b) Solve $2x = 9$ **[1 mark]**

Answer $x =$ ....

(c) Calculate $a^2 - 3a$ when $a = 5$ **[2 marks]**

Answer ....

2. (a) Which of these is **both** a multiple of 3 and a **square** number? Circle your answer. **[1 mark]**

1 | 2 | 3 | 9 | 18 | 27

(b) Which of these is **not** a prime number? Circle your answer. **[1 mark]**

2 | 3 | 7 | 13 | 27 | 31

3. Complete the diagram so that it makes the net of a cube. **[1 mark]**

4. These were the lowest recorded temperatures in three British cities last year:

Manchester: −8.3°C	Edinburgh: 3.2°C colder than Manchester
London: 3.6°C warmer than Manchester	

(a) What was the lowest recorded temperature in Edinburgh last year? **[1 mark]**

Answer .... °C

(b) What was the lowest recorded temperature in London last year? **[1 mark]**

Answer .... °C

5. **(a)** Write $\frac{3}{8}$ as a decimal. **[1 mark]**

Answer ....

(b) Work out $\frac{13}{16} - \frac{1}{4}$ **[2 marks]**

Answer ....

(c) Work out $\frac{3}{4}$ of 3 **[2 marks]**

Answer ....

6.

106°

47°

29°

Is it possible to draw this triangle accurately? Give a reason for your answer. **[2 marks]**

....

....

....

7. There are four blue counters, three white counters and two red counters in a bag.

Sandeep is going to take one counter and keep it. Then Colin is going to take one.

(a) Find the probability that Sandeep chooses a white counter. **[1 mark]**

Answer ....

Sandeep does choose a white counter.

(b) Find the probability that Colin chooses a white counter. **[1 mark]**

Answer ....

8. Libby is thinking of a number, n.

She subtracts 5 from her number and then multiplies her answer by 3.

Write an algebraic expression to show this. **[2 marks]**

Answer ....

9. Becky and Nate share £420 in the ratio 3 : 4

(a) What fraction of the money does Becky get? **[1 mark]**

Answer ....

(b) How much money does Nate get? **[3 marks]**

Answer £ ....

10. Wayne is buying ingredients for a meal.

(a) Complete the bill for Wayne's shopping. **[3 marks]**

Item	Price	Quantity	Total
Side of salmon	£9.55	1	£9.55
Onions	£0.36	5	....
Rice	....	2	£2.36
Garlic	£0.17	....	£0.51
White wine	£4.49	2	£8.98
		Total	....

Wayne has a voucher which gives him a 15% discount on his total bill.

(b) How much does Wayne pay for his shopping? **[3 marks]**

Answer £ ....

11. A martial arts club has 53 members.

- 17 members are female
- 34 members have never won a medal
- 8 of the males have won a medal

(a) Complete the two-way table to show this information. **[2 marks]**

	Won a medal	Never won a medal	Total
Male	8	....	....
Female	....	....	17
Total	....	....	53

(b) How many females have won a medal? **[1 mark]**

Answer ....

12. Calculate the value of $2^{-3} \times \sqrt[5]{32}$ **[2 marks]**

Answer ....

13. A pig farm has three different types of rare-breed pig.

Complete the table and pie chart accurately for the following information. The diagram is drawn accurately. **[3 marks]**

Berkshire	Middle-white	British lop
23	32	....

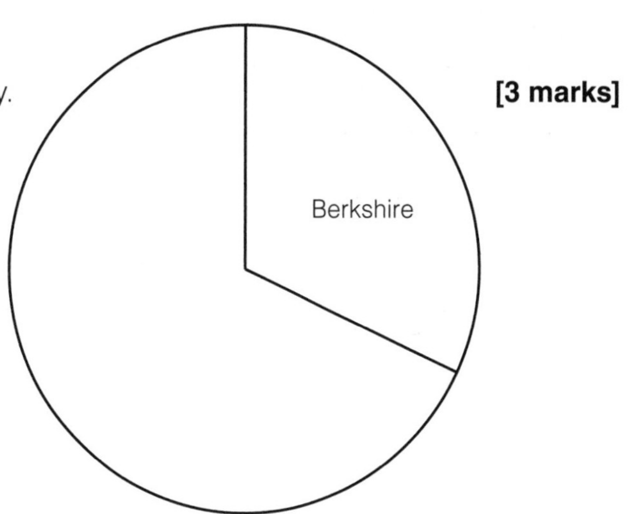

14. Triangle A is the larger triangle. Triangle B is inside triangle A.

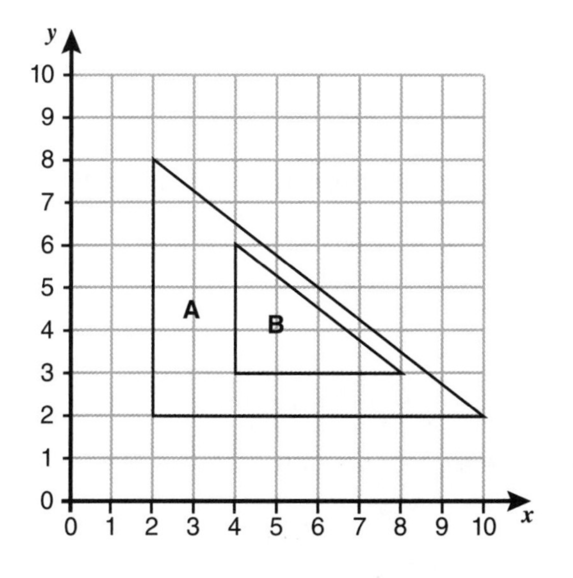

Describe the single transformation which maps triangle A onto triangle B. **[3 marks]**

....

....

15. The table shows the number of British tourists who visited Chile from 2006 to 2015.

Year	2006	2007	2008	2009	2010
British tourists (nearest 1000)	61000	68000	72000	75000	42000
Year	**2011**	**2012**	**2013**	**2014**	**2015**
British tourists (nearest 1000)	47000	55000	58000	63000	71000

(a) Complete the time-series graph to show this information. **[1 mark]**

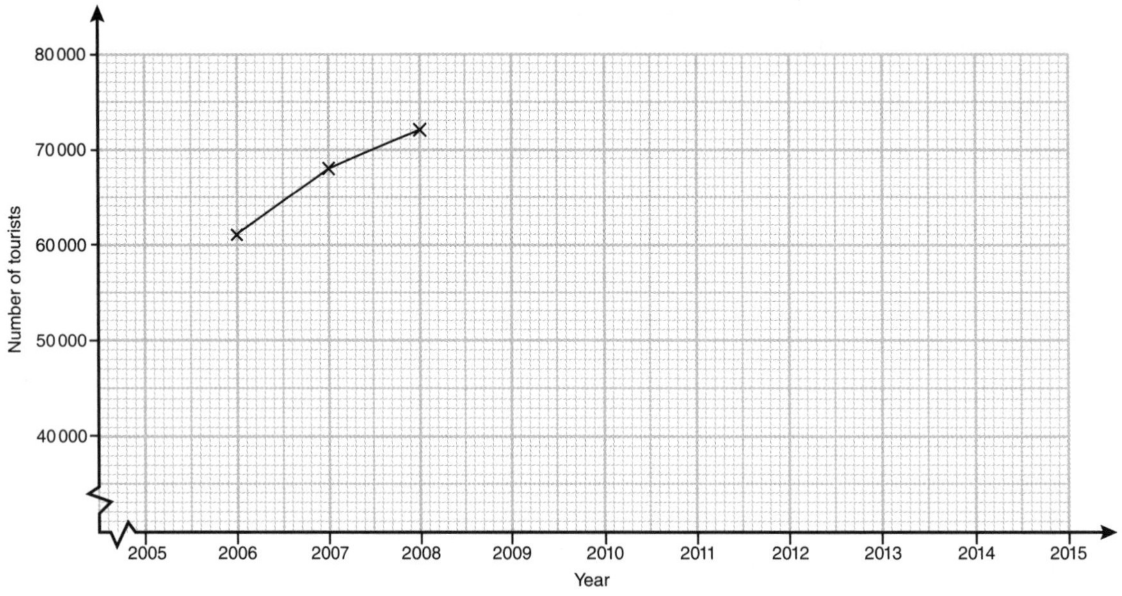

(b) Describe how the number of British tourists to Chile changed between 2006 and 2015. **[2 marks]**

....

....

....

16. Tick whether each statement is **true** or **false**. Give a reason for your answer.

(a) The value of x^3 is **always** positive. **[1 mark]**

☐ True ☐ False

Reason ....

....

(b) The value of x^2 is **always** positive. **[1 mark]**

☐ True ☐ False

Reason ....

....

(c) The value of x^2 is **never** equal to the value of $2x$. **[1 mark]**

☐ True ☐ False

Reason ....

....

17. (a) Solve the inequality $-6 \leqslant 2x - 3 < 3$ **[2 marks]**

Answer ....

(b) Write down all the integers which satisfy $-6 \leqslant 2x - 3 < 3$ **[1 mark]**

Answer ....

18. Clive wants to find out how often people visit the library. He designs this question for a survey:

How many times do you visit the library? Tick a box.			
☐	☐	☐	☐
1 to 2 times	*3 to 4 times*	*4 to 5 times*	*6 or more times*

(a) Give one criticism of Clive's question. **[1 mark]**

....

....

(b) Design a better question which Clive could use. **[2 marks]**

19. Rachel is the exams officer at a school. She finds that 38 out of 203 pupils do not bring a calculator to a maths exam.

(a) Out of 48 752 students taking the maths exam in England, estimate how many will not bring a calculator to the exam. **[2 marks]**

Answer .... students

(b) Have you made an over-estimate or an under-estimate?

State any assumptions you have made. **[2 marks]**

....

....

....

20. Violet wants to invest £2500 for three years. She looks at two different options:

Premium bonds
For every £1000 you invest, you are likely to receive a cheque for £15 per year. (This is not compound interest).

Local bank (compound interest)
• 1% interest for the first year • 1.5% interest for the second year • 2% interest for the third year

Which option is likely to give the most money if she invests £2500 for three years? **[5 marks]**

Answer ....

21. Shape $ABCD$ is a quadrilateral such that:

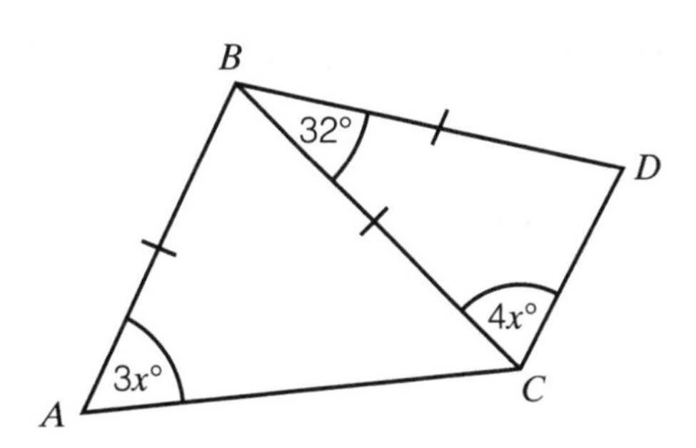

$AB = BC = BD$

$BAC = 3x°$

$BCD = 4x°$

$CBD = 32°$

Find the value of the missing angle ABC. **[4 marks]**

Answer .... °

22. Matt took part in a 10 km run. He burns 133 calories for every mile he runs.

The run took him 1 hour and 15 minutes to complete. His usual average speed on level tarmac is 9 miles per hour (mph).

(a) How many calories did Matt burn on the run? **[4 marks]**

Answer .... calories

(b) Select which of these options, A, B, C or D, best describes the 10 km run.
Give a reason for your answer.

A: Streets and pavements around town with no hills

B: A cross-country run with muddy, rocky paths and steep inclines

C: Several laps around a running track

D: A run along a sandy beach **[3 marks]**

Answer ....

Reason ....

....

23. Solve the simultaneous equations

$4a - b = 7$

$6a - 2b = 9$ **[4 marks]**

Answer $a =$ .... and $b =$ ....

24. The equilateral triangle, with side length 2, has been split into two congruent, right-angled triangles.

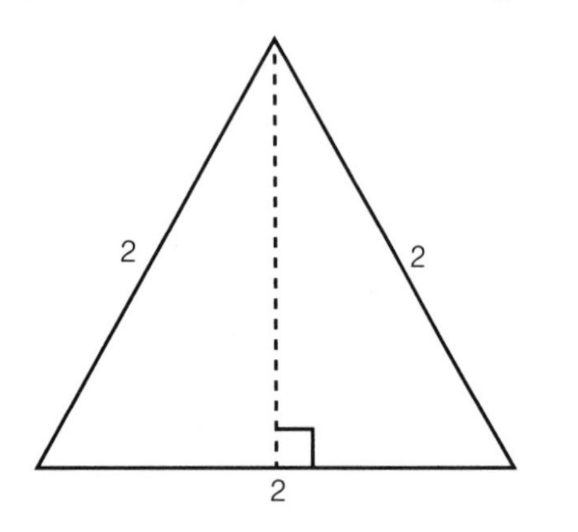

Use this diagram to prove that $\sin 60° = \frac{\sqrt{3}}{2}$ **[3 marks]**

END OF QUESTIONS

END OF PRACTICE EXAM 2

You are encouraged to show your working out, as you may be awarded marks for method even if your final answer is wrong. Full marks can be awarded where a correct answer is given without working being shown, but if a question asks for working out, you must show it to gain full marks. If you use a correct method that is not shown in the mark scheme below, you would still gain full credit for it.

PRACTICE EXAM 1

Paper 1 (pages 3–13)

1. 0.75

2. $a + 2b$ **[1 mark for each correct term]**

Collect the like terms.

3. 4 + 6 = 10

4. 12 : 15 **[1]**

4 : 5 **[1]**

5. **(a)** B **(b)** D

6. $1 - \frac{7}{20}$ **[1]** $= \frac{13}{20}$ **[1]**

7. **(a)** 2 is a prime number **(b)** 1 is a factor of 16

(c) 9 – 1 = 8 or 16 – 4 = 12, etc.

8. 20 **[2]**

[1 mark for an answer of 10 or 1 mark for four cards seen in order with 14 and 18 in the middle]

The two middle numbers need to be 14 and 18 to give median of 16. 12 + a range of 8 gives 20.

9. Factors of 24: 1, 2, 3, 4, 6, 8, 12 and 24 **[2]**

[1 mark for at least six correct]

2 + 8 + 6 = 16 **[1 mark for adding any three factors]**

2, 6 and 8 or 1, 3 and 12 **[1]**

10. $4x - 8 = 2$ **[1]**

$4x = 10$

$x = 2.5$ **[1]**

Expand the bracket, add 8 to both sides and then divide by 4.

11. 6.40 + 6.40 + 8.25 + 3.50 + 2 + 2 + 1.50 + 1.50 **[2]**

= £31.55 **[1]**

[1 mark for at least six out of eight correct values added]

12. (Air Jump, 3G Swing, High Ropes); (Air Jump, High Ropes, 3G Swing);

(3G Swing, Air Jump, High Ropes); (3G Swing, High Ropes, Air Jump);

(High Ropes, Air Jump, 3G Swing); (High Ropes, 3G Swing, Air Jump).

6 **[3] [2 marks for all six combinations shown; 1 mark for at least four shown]**

13. **(a)** 21 **[2] [1 mark for a triangle with 21 circles drawn]**

(b) The sequence goes: Odd number, Odd number, Even number, Even number, Odd number, Odd number, Even number, Even number…

Or Add consecutive odd and even numbers to get the next pattern.

14. Straight line graph passing through points (–2, –5), (–1, –3), (0, –1), (1, 1), (2, 3), (3, 5) **[3]**

[2 marks for at least four points correctly plotted; 1 mark for at least one point correctly plotted]

Construct a table of values for x and y.

15. **(a)** 15% of 1200 = 180 students **[1]**

1200 + 180 = 1380 students **[1]**

(b) 1380 – 138 (or subtract 10% from answer to part (a)) **[1]**

= 1242 students **[1]**

16. Shop A: 140p ÷ 5 = 28p **[1]**

Shop B: 116p ÷ 4 = 29p **[1]**

Shop A gives the best value for money. **[1]**

Use the bus-stop method for division.

17. **(a)** 140 cm ÷ 25 cm **[1]**

5.6 rounds to 5 complete tiles **[1]**

Convert to the same units.

(b) 0.6 × 25 **[1]** = 15 cm **[1]**

18. $\frac{21}{35} + \frac{10}{35}$ **[1]** $= \frac{31}{35}$ **[1]**

19. Length scale factor = 2 **[1]**

Area scale factor = $2^2 = 4$ **[1]**

No **[1]** Christopher's photograph is four times the area of Liam's photograph. **[1]**

Draw two similar rectangles with a scale factor of 2 and calculate the area of each of them.

20. 200 ÷ 5 **[1]** = £40 per share **[1]**

Charlotte = £160 (4 × £40), Clarisse = £240 (6 × £40), Mathilda = £440 (11 × £40) **[1]**

Mathilda gets five more shares than Clarisse.

21. $BCD = 68°$ (alternate angles) **[1]**

$CBD = 56°$ (isosceles triangle) **[1]**

$ABD = 124°$ (straight line), $DAB = 34°$ (angles in a triangle) **[1]**

$x = 360 - 68 - 34 = 258°$ (angles around a point) **[1]**

22. **(a)** Either $\frac{2}{7}$ (Tim's results), $\frac{11}{31}$ (Amanda's results) or $\frac{13}{38}$ (combined) **[1 mark for each up to a maximum of 2]**

(b) Either Amanda's or combined **[1]** depending on answer to part (a).

Reason: More days being recorded will give more accurate estimate **[1]**

23. **(a)** $2x - 4 \leqslant 7$ **[1 mark for correct first step]**

$2x \leqslant 11$

$x \leqslant 5.5$ **[1]**

Solve the inequality like an ordinary equation.

(b)

0 1 2 3 4 5 6 7 8 9 10

[1 mark for correct line segment; 1 mark for shaded circle. 2 marks for the correct diagram based upon incorrect part (a)]

24. 610 000 + 42 000 = 652 000 **[1]**

6.52×10^5 **[1]**

Answers

25. $8x^2 - 4x + 12x - 6$ **[1]**
$8x^2 + 8x - 6$ **[1]**

26. £180000 = 120% **[1]**
£1500 = 1%
£150000 = 100% **[1]**

27. $10 \div 4 = 2.5$ **[1]**
$AB = 7.5 \div 2.5 = 3$ cm **[1]**
$AC = 5$ cm **[1]**
$AE = 5 \times 2.5 = 12.5$ cm **[1]**
$CE = 12.5 - 5 = 7.5$ cm **[1]**

AC is 5 cm since ABC is a (3, 4, 5) triangle, using Pythagoras.

28. (a)

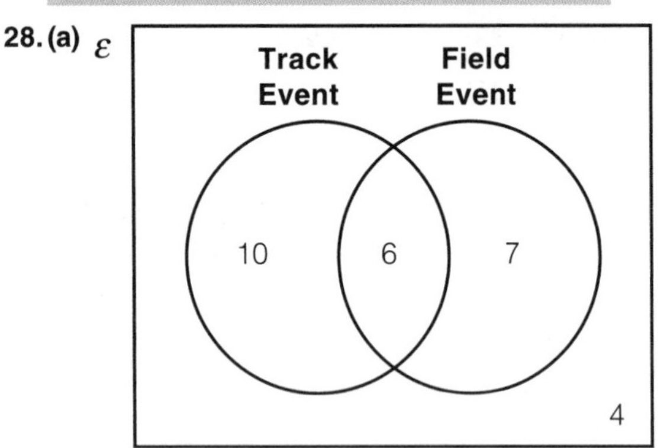

[2 marks if fully correct; 1 mark for at least two correct values]

(b) $\frac{6}{27}$ **[2] [1 mark for either 6 or 27 identified]**

Paper 2 (pages 14–25)

1. **(a)** 10% = £6 20% = £12
(b) $\frac{40}{100} = \frac{4}{10} = \frac{2}{5}$ **(c)** 0.38

2. **(a)** 108–112° **(b)** 6.6–6.8 cm

3. **(a)** C A rhombus has four equal sides.
(b) Adjacent sides are equal; One line of symmetry; Diagonals cross at 90° **[Any one for 1 mark]**

4. Exactly two tiles labelled A; three tiles labelled B; one tile labelled with any letter apart from A, B or C. **[2 marks if fully correct; 1 mark for one of these satisfied]**

5. **(a)** True **(b)** False **(c)** False **(d)** True

6. **Any suitable answer, e.g.:**

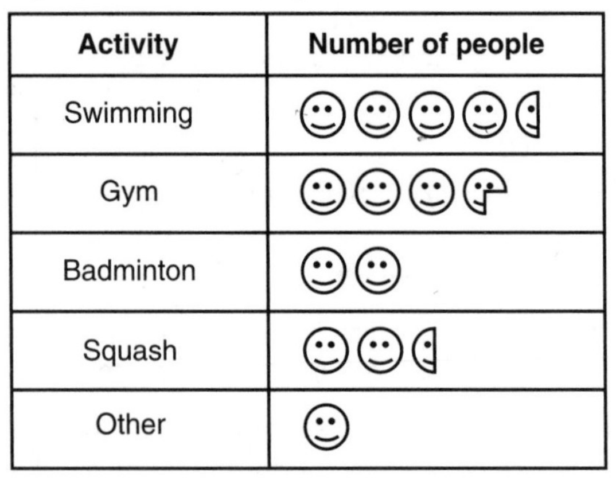

Activity	Number of people
Swimming	
Gym	
Badminton	
Squash	
Other	

Key = 4 people

[3 marks if fully complete with a key; lose a mark for each error up to a maximum of 3]

7. **(a)** $A = 3$, $B = 4$ (or any multiples of 3 and 4).
(b) $4.5 \times 4 = 18$ **[1]**
$18 \div 3 = 6$ **[1]**

Reverse the operations in part (a).

8. **(a)** $xy + 7$ **[2] (accept $x \times y + 7$)**
[1 mark for xy or $x \times y$ seen]
(b) $xy + 7 = 39$ **[1]**
$8x + 7 = 39$ **[1]**
$8x = 32$
$x = 4$ boxes **[1]** Use part (a) to write an equation.

9. $3n + 2$ **[2] [1 for $3n$ seen]**

$3n$ comes from the difference of 3.

10. **(a)** Distance = 8.8 cm **(accept 8.6–9.0 cm)**
8.8×0.25 **[1]** = 2.2 km **[1] (accept 2.15–2.25 km)**
(b) 45 minutes = 0.75 hours **[1]**
$\frac{2.2 \text{ km}}{0.75 \text{ h}}$ **[1]** = $2.9\dot{3}$ km/h **[1] (accept follow-through answers from part (a))**

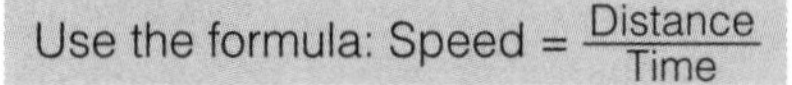
Use the formula: Speed = $\frac{\text{Distance}}{\text{Time}}$

11. **(a)**

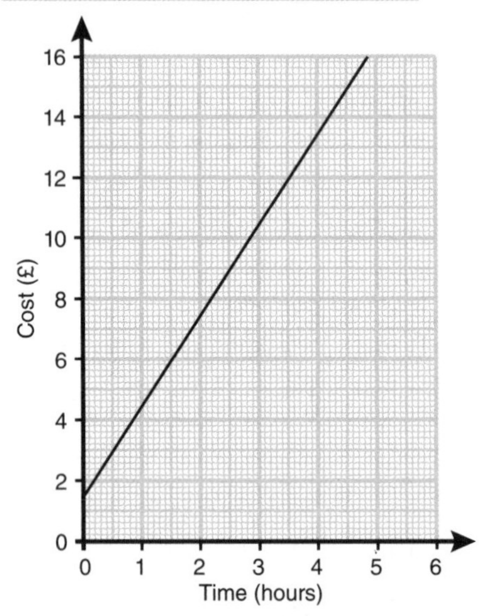

[3 marks if fully correct from 0–4 hrs; 2 marks for a partially complete graph from 0–4 hrs; 1 mark for at least one point that would lie on the line]

(b) 3.2 hours **[1] (accept 3.1 to 3.3 hours)**
3 hours and 12 minutes **[1] (accept 3 hours 6 minutes to 3 hours 18 minutes)**

0.1 hours = 6 minutes (60 ÷ 10)

12. $v - u = at$ **[1]** $\frac{v - u}{t} = a$ **[1]**

Subtract u from both sides, then divide by t.

13. **(a)**

1st	2nd	3rd
R	J	A
R	A	J
J	A	R
J	R	A
A	R	J
A	J	R

[2 marks if fully correct; 1 mark for at least three combinations]

(b) 3 identified **[1]** $\frac{3}{6}$ or $\frac{1}{2}$ **[1]**

14.

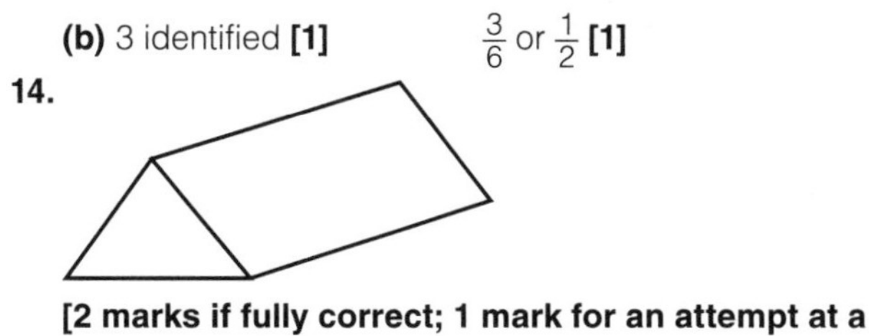

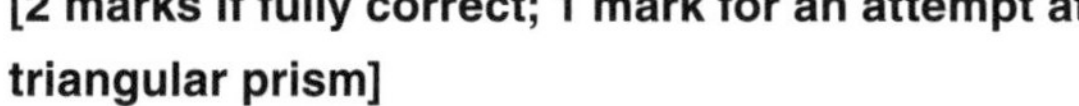

[2 marks if fully correct; 1 mark for an attempt at a triangular prism]

15. (a)

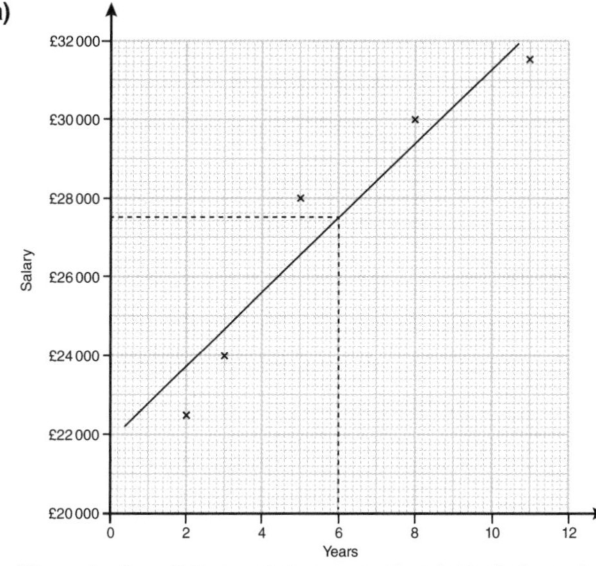

[2 marks for all five points correctly plotted; 1 mark for one error only]

(b) Answer in range £27 000 to £28 000

(c) It is outside the given range of the data.

16. $\pi \times 2 \times 25$ **[1]** = 157.079… m **[1]** **(accept 157 m)**

157.079… × $\frac{60}{360}$ + 2 × 25 **[1]** = 76.2 m **[1]**

> Calculate the circumference of the whole circle using C = π × 2 × r

17. (a) $4x(3x - 1)$

(b) $12x^3y^2$

(c) $3 \times (-2)^2 - (-2) = 14$

18. (a) $\frac{4^3}{\sqrt{100} - 6}$ **[1]** = $\frac{64}{4}$ = 16 **[1]**

> Round all the values to 1 significant figure.

(b) $\frac{59.319}{3.678\,699\,101}$ = 16.124 993 75

> Do not round at any point.

(c) 16.1

19. (a) £450 × 1.285 **[1]** = €578.25 **[1]**

(b) France: 36 ÷ 1.285 = £28.02 **[1]**

£28.02 – £27 = £1.02 **[1]**

20. Beans: 14, 28, 42, **56**, 70, 84

Soup: 8, 16, 24, 32, 40, 48, **56**, 64

[1 mark for listing multiples of 14 and 8; 1 mark for finding a common multiple of 8 and 14, e.g. 56, 112, etc.]

4 boxes of beans and 7 boxes of soup (or any multiples of 4 and 7) **[1]**

21. 500 g + 1000 g + 1500 g = 3000 g

3000 g ÷ 2170 cm³ **[1]** = 1.38 g/cm³ **[1]**

> Convert all amounts into grams and cm³.

22. £152 000 × $(1.02)^1$ **[1]** × $(1.025)^2$ **[1]** = £162 888.90 **[1]**

23. (a) 1 **(b)** $\frac{1}{2^5}$ **[1]** = $\frac{1}{32}$ **[1]**

24. Equation 1: $15x + 10y = 25$

Equation 2: $15x - 21y = 63.75$

Equation 1 – Equation 2: $31y = -38.75$ **[1]**

$y = -1.25$

$3x + 2(-1.25) = 5$ **[1]**

$3x = 7.5$

$x = 2.5$, $y = -1.25$ **[1]**

> Substitute the y-value back into one of the original equations.

25. $\sin 63° = \frac{18}{RS}$ **[1]**

$RS = \frac{18}{\sin 63°}$ **[1]**

RS = 20.2 cm **[1]**

Paper 3 (pages 26–37)

1. **(a)** 3700 **(b)** 6 hundred **(accept 600 or hundreds)**

   (c) 4

2. **(a)** Factors of 12: 1, 2, 3, 4, 6 and 12 **[1 mark for listing at least four factors]**

   2 or 3 **[1]**

   (b) Any one of the following: 49, 81, 121, 169, etc.

   > Find the odd square numbers by squaring odd digits e.g. $7^2 = 49$

3. **(a)** (2, –4) **(b)** (–0.5, 2) **[1 mark for each correct value]**

4. $8a^2 - 4ab$ **[1 mark for each correct term]**

   > Multiply the term outside the bracket by each term inside the bracket.

5. 85.2 kg **[1]** identified from the scale

   85.2 kg + 1.4 kg + 0.8 kg **[1]** = 87.4 kg **[1]**

6. **(a)** $\frac{25 - 15}{25} = \frac{10}{25}$ **[1]** = $\frac{2}{5}$ **[1]**

   (b) $\frac{16}{25} \times 100$ **[1]** = 64% **[1]**

7. **(a)** 1625.22 kg ÷ 12 = 135.435 kg **[1]**

   135.435 × 15 = 2031.525 kg **[1]**

   > Find the weight of one washing machine first.

   (b) 2600 kg ÷ 135.435 **[1]** = 19.1974…

   19 **[1]**

   > Convert 2.6 tonnes to 2600 kg by multiplying by 1000.

8. 3 × (4 + 5) ÷ 6 = 4.5

   [2 marks if fully correct; 1 mark for correct symbols but missing brackets]

9. **(a)** $25n + 85t$ **[1]**

   $C = 25n + 85t$ **[1]**

   (b) $C = 25 \times 17 + 85 \times 2$ **[1]** = £595 **[1]**

   (c) $560 = 25n + 85$ **[1]**

   $475 = 25n$

   $n = 19$ **[1]**

   > Subtract 85 from both sides. Then divide both sides by 25.

10. 655 ÷ 2.54 **[1]** = 257.874…

    258 inches **[1]**

11. $250 \times 6.4 \div 100$ **[1]** = 16 items **[1]**

Expected number of returns = Probability of return × Total number of items

12. (a) Micah's age: $2x$ **[1]**

Kayley's age: $2x + 4$ years old **[1]**

(b) $x + 2x + 2x + 4 = 19$ **[1]**

$5x = 15$

$x = 3$ **[1]**

Micah's age: $2 \times 3 = 6$ years old **[1]**

Write an expression for the sum of the three ages in terms of x, and equate it to 19. Solve the equation to find x. Micah's age is $2x$.

13. 100 m in 10 seconds = 10 m/s **[1]**

$10 \times 60 \times 60 = 36000$ **[1]** m/h

$36000 \div 1000 = 36$ km/h **[1]**

14. Missing lengths: 6 m and 2.5 m **[1 mark for one missing length found]**

$8 \times 4 + 2 \times 2.5$ or $2 \times 6.5 + 6 \times 4$ **[1]**

Total area = 37 m^2 **[1]**

£3.99 × 37 = £147.63 **[1]**

Split the shape into two rectangles.

15. 0.035×5450 = £190.75 **[1]**

£190.75 × 4 = £763 **[1]**

16. 3 cm, 4 cm, 5 cm **[3 marks if fully correct, where 5 cm is the hypotenuse; 1 mark if the product of the two shorter sides is 12; 1 mark if the sum of the three sides is 12 cm]**

Area of a triangle = $\frac{\text{Base} \times \text{Height}}{2}$. Recognise the Pythagorean triple ($3^2 + 4^2 = 5^2$)

17. (a) $1260 = 2 \times 2 \times 3 \times 3 \times 5 \times 7$ **[2]** $= 2^2 \times 3^2 \times 5 \times 7$ **[1]**

[1 mark for first two correct steps using a prime factor tree or repeated division]

(b) 1050 = **2** × **3** × **5** × 5 × **7**

1260 = **2** × 2 × **3** × 3 × **5** × **7**

HCF = $2 \times 3 \times 5 \times 7$ **[1]** = 210 **[1]**

HCF is the product of the identical prime factors from each list.

18. C : D : J

$4 \times \frac{1}{3}$J : $\frac{1}{3}$J : J **[1]**

4 : 1 : 3 **[1]**

If Julien is one unit, Del is one-third and Carolyn is four lots of one-third.

19.

Time	10 s	20 s	30 s	40 s	**45 s**	50 s
Nat's distance	20 m	60 m	100 m	140 m	**160 m**	180 m
Tim's distance	0 m	0 m	40 m	120 m	**160 m**	200 m

[3 marks for full working leading to 160 m or 45 seconds; 2 marks for at least two areas calculated for each person; 1 mark for at least one area calculated]

The race was 160 m long **[1]**.

Work out the area under each graph to find the distance travelled.

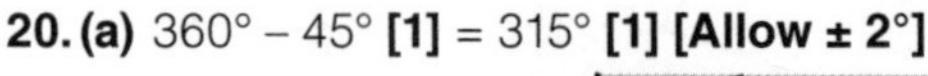

20. (a) 360° – 45° **[1]** = 315° **[1]** **[Allow ± 2°]**

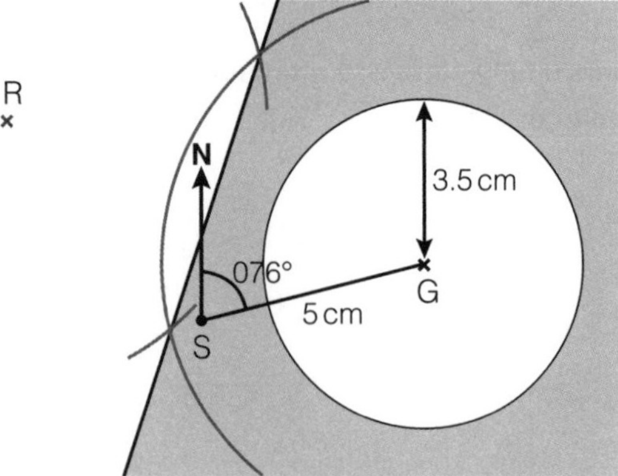

(b) G marked at a bearing of 076° **[1]** where SG = 5 cm **[1]**

(c) Perpendicular bisector of RG drawn.

A circle, radius 3.5 cm, at G. **[1 mark for one correct construction]**

Correct region shaded **[1]**

21. Jill's mean: $(50 \times 3 + 150 \times 8 + 250 \times 6 + 350 \times 7)$ **[1]**

$\div 24$ = £220.83 **[1]**

David's mean: $(50 \times 2 + 150 \times 4 + 250 \times 8 + 350 \times 6)$ **[1]**

$\div 20$ = £240 **[1]**

David earned the most per painting **[1]**

Multiply the midpoints of each interval by the frequencies and divide by the total frequency.

22. (a) E **(b)** C **(c)** F

23. (a) First branch: 0.25 **[1]**; Second branch: 0.35, 0.65, 0.35 **[1]**

(b) P(Late, Late) = 0.25×0.35 **[1]** = 0.0875 **[1]**

PRACTICE EXAM 2

Paper 1 (pages 38–49)

1. (a) 4 or 10 **(b)** 10 or 4 **(c)** 5

The mode is the number that occurs the most often.

2. (a) 13 **(b)** 8

$2 \times 2 \times 2 = 8$

3. (a) $2a - b$ **[1 mark for each correct term]**

(b) $20xy$

4. 20% = 40 or 1% = 2 **[1]**

21% = 40 + 2 = 42 **[1]**

5. $\frac{35}{60}$ **[1]** $= \frac{7}{12}$ **[1]**

6. $\frac{12}{36} = \frac{1}{3} = 0.333333\ldots$, 32.5% = 0.325

$\frac{8}{25} = \frac{32}{100} = 0.32$ **[1 mark for at least one correct]**

$\frac{8}{25}$, 32.5%, $\frac{12}{36}$, 0.43 **[1]**

Convert all values to decimal equivalents.

7. Possibility 1: £1, 20p, 20p, 10p, 10p, 5p **[1]**

Possibility 2: 50p, 50p, 20p, 20p, 20p, 5p **[1]**

8. (a) = **(b)** $>$

9. **(a)** 0.35 kg = 350 g **[1]**

$\frac{350}{12}$ **[1]** = 29 **[1]** with some gold remaining

(b) 350 – 29 × 12 = 350 – 348 = 2 g

> Use long division, or look for some multiples of 12 close to 350, e.g. 12 × 30 = 360

10. No **[1]** with correct working $\frac{5}{6} - \frac{2}{6} = \frac{3}{6} = \frac{1}{2}$ **[1]**

Or No **[1]** with reasons: $\frac{5}{6} - \frac{1}{3}$ is less than 1 and $\frac{4}{3}$ is greater than 1 **[1]**

> Draw a sketch to show that the answer should be less than 1.

11. $\frac{-2 - -3}{-2} = \frac{1}{-2} = -\frac{1}{2}$

$\frac{-2 \times -3}{-2} = \frac{6}{-2} = -3$

$\frac{-3 - -2}{-2} = \frac{-1}{-2} = \frac{1}{2}$

[2 marks for all three correctly evaluated; 1 mark for at least one correct]

$\frac{b - a}{a}$ **[1]** gives a positive value.

12. $5n - 3 = 45$ **[1]**

$5n = 48$

No, because 48 is not a multiple of 5. **[1]**

13. **(a)** 180 ÷ (1 + 3 + 2) **[1]** = 30° **[1]**

$x = 30°$, $y = 90°$ and $z = 60°$ **[1]**

> Alternatively write, and solve, the equation $x + 3x + 2x = 180°$

(b) A right-angled triangle **(accept scalene triangle)**

14. **(a)**

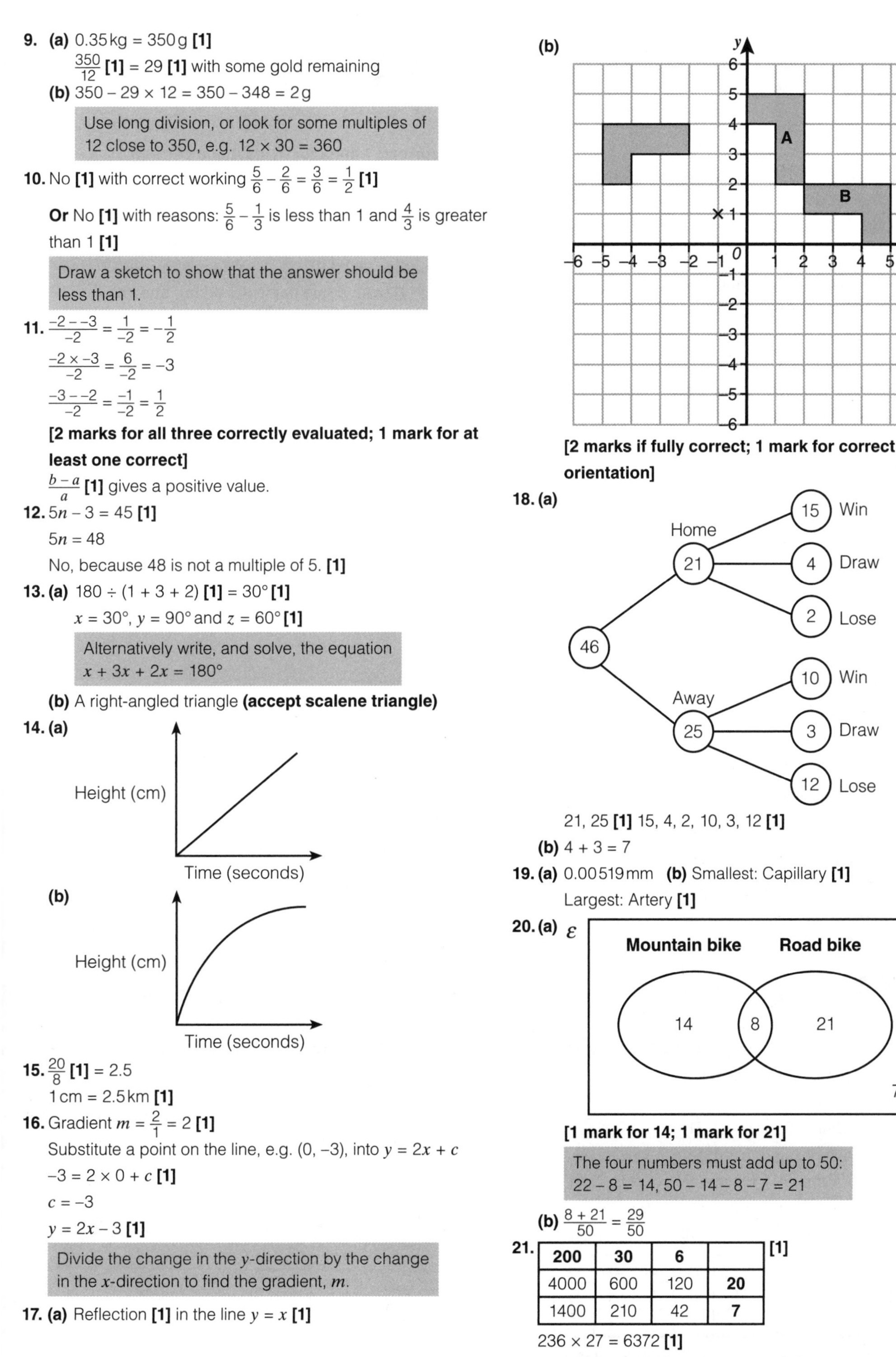

(b)

15. $\frac{20}{8}$ **[1]** = 2.5

1 cm = 2.5 km **[1]**

16. Gradient $m = \frac{2}{1} = 2$ **[1]**

Substitute a point on the line, e.g. (0, –3), into $y = 2x + c$

$-3 = 2 \times 0 + c$ **[1]**

$c = -3$

$y = 2x - 3$ **[1]**

> Divide the change in the y-direction by the change in the x-direction to find the gradient, m.

17. **(a)** Reflection **[1]** in the line $y = x$ **[1]**

(b)

[2 marks if fully correct; 1 mark for correct orientation]

18. **(a)**

21, 25 **[1]** 15, 4, 2, 10, 3, 12 **[1]**

(b) 4 + 3 = 7

19. **(a)** 0.00519 mm **(b)** Smallest: Capillary **[1]**

Largest: Artery **[1]**

20. **(a)**

[1 mark for 14; 1 mark for 21]

> The four numbers must add up to 50: 22 – 8 = 14, 50 – 14 – 8 – 7 = 21

(b) $\frac{8 + 21}{50} = \frac{29}{50}$

21. **[1]**

200	**30**	**6**	
4000	600	120	**20**
1400	210	42	**7**

236 × 27 = 6372 **[1]**

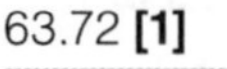

63.72 **[1]**

> Calculate without the decimal point but remember to restore it in the answer.

22. $5x - 20 = 3x - 15$ **[1]**

$2x = 5$ **[1]**

$x = 2.5$ **[1]**

23. $\frac{3}{5} \times \frac{15}{4}$ **[1]** $= \frac{45}{20}$ **[1]** $= 2\frac{5}{20} = 2\frac{1}{4}$ **[1]**

24. (a) $\frac{8}{30}$ or $\frac{4}{15}$

(b) $\frac{3}{30} \times \frac{5}{30}$ **[1]** $= \frac{1}{10} \times \frac{1}{6} = \frac{1}{60}$ **[1]**

Use the AND rule.

25. $\frac{350 - 280}{350}$ **[1]** $= \frac{70}{350} = \frac{1}{5} = 20\%$ **[1]**

26. $51 = \frac{\text{distance}}{0.666...}$ **[1]**, $d = 51 \times 0.666... = 34$ miles **[1]** from Manchester to Knutsford services.

$66 - 34 = 32$ miles from Knutsford services to Stoke-on-Trent

Speed $= \frac{32}{0.5} = 64$ mph **[1]**

Use speed $= \frac{\text{distance}}{\text{time}}$, converting minutes into hours.

27. Triangle ABC is congruent to triangle GHI. **[1]**

Using ASA, both triangles have a 5 cm side between the two angles 30° and 70°, proving that all three sides are identical. **[1]**

In triangle ABC, use the fact that angles in a triangle add up to 180°.

28. Arc length $= \frac{2 \times \pi \times 2x}{4}$ **[1]** $= x\pi$

Perimeter $= x\pi + 2x + 2x$

$= 4x + x\pi$ or $x(4 + \pi)$ **[1]**

Use the formula: Circumference $= 2\pi r$

29. $-2 + 2 \times r = 4$

$q + 2 \times 8 = 21$ **[1]**

$r = 3$ and $q = 5$ **[1]**

30. $(x - 12)(x + 3) = 0$ **[1]**

$x - 12 = 0$ or $x + 3 = 0$

$x = 12$ and $x = -3$ **[1]**

31. $877.5 \leqslant m < 882.5$

[1 mark for correct numbers; 1 mark for correct symbols]

Error interval is half of 5 miles above and below 880 miles.

Paper 2 (pages 50–61)

1. (a) 34 600 **(b)** 428 000

2. 0.083, 0.8, 0.803, 0.83, 0.883 **[2 marks if fully correct; 1 mark for three in correct order]**

3. (a) Table completed with these frequencies:

White = 5, Blue = 9, Black = 7 **[1]**

Bar chart completed with these frequencies:

Silver = 10, Green = 4, Maroon = 2, Other = 3 **[1]**

(b) $(8 + 5 + 10 + 9 + 4 + 7 + 2 + 3) + 8 - 5 = 51$

4. (a) 256 **(b)** 23 **(c)** 5

5. (a)

Minutes	**A**	**A**	A	A	B	B	B	B
Texts	**A**	**A**	B	B	A	A	B	B
Data	**A**	**B**	A	B	A	B	A	B

[2 marks if fully correct; 1 mark for at least four more columns correct]

(b) $\frac{1}{8}$

6. $41.40 \div 225$ **[1]** = £0.184 or 18.4p **[1 (with correct units)]**

Convert the hours to minutes first.

7. (a) Triangular prism

(b) 5 faces, 9 edges and 6 vertices **[2 marks if fully correct; 1 mark for at least one correct]**

8. (a) 79°F

Draw a vertical line at 26 up to the graph, then read off the vertical axis.

(b) −20°C = −4°F **[1]**

No **[1] (with some correct working shown)**

9. Translation **[1]** with the vector $\begin{pmatrix} 2 \\ -4 \end{pmatrix}$ **[1]**

10. $\frac{1}{6}$ or $\frac{2}{11}$ **[1]** seen

$\frac{1}{6} = \frac{2}{12} < \frac{2}{11}$ **[1]**

The spinner gives the best chance since $\frac{2}{11} > \frac{2}{12}$ **[1]**

11. Booties: £3.59 × 2 = £7.18 **[1]**

Hair-Care: £3.79 × 3 = £11.37

0.4 × £11.37 = £4.548 **[1]**

= 11.37 − 4.548 = 6.822 = £6.82 **[1]**

Hair-Care gives the best value **[1]**

Calculate the cost of two bottles at Booties, since you get the third for free.

12. Self-raising flour: 240 g Caster sugar: 240 g

Rolled oats: 192 g Butter: 240 g

Syrup: 24 g

[3 marks if fully correct; 2 marks with only one error; 1 mark for at least one correct or 1.6 seen]

Multiply each of the original amounts by 1.6, since $40 \div 25 = 1.6$

13. Adder: Median $= \frac{35 + 1}{2} =$ 18th position **[1]**

Grass snake: Median $= \frac{39 + 1}{2}$ 20th position **[1]**

No, median class interval: Adder ($60 < l \leqslant 65$); Grass snake ($65 < l \leqslant 70$) **[1]**

The 18th adder lies in the 3rd class interval; the 20th grass snake lies in the 4th class interval.

To find the median position, use the formula: $\frac{\text{Total frequency} + 1}{2}$

14. $\frac{4}{5}$ of 30 = 24 **[1]** children

$\frac{1}{6}$ of 24 = 4 **[1]** children have two lessons per week

4 × 2 × £16 **[1]** + 20 × £16 + 6 × £21 **[1]** = £574 **[1]**

15. $\frac{2}{60}$ **[1]** $= \frac{1}{30}$ fish marked

$\frac{1}{30} \times \frac{40}{40} = \frac{40}{1200}$ **[1]** fish are marked

1200 fish **[1]** in the lake

Since 40 fish are marked, use equivalent fractions to estimate the number of fish in the lake.

16. (a) $2x + 3 = 10$ **[1]**

$x = 3.5$ **[1]**

(b) $v^2 - 2as = u^2$ **[1]**

$u = \sqrt{v^2 - 2as}$ **[1]**

17. $\frac{1587392 - 1372658}{1372658} = \frac{214734}{1372658} \times 100$ **[1]**

15.6% **[1]**

18. $2 \div (15 - 11) = 0.5$ **[1]** sweets per share

15×0.5 **[1]** $= 7\frac{1}{2}$ sweets **[1]**

19. $0.\dot{6} \times 162 = 108$ **[1]** go to theme park

$\frac{8}{27} \times 162 = 48$ **[1]** go ice-skating

$162 - 108 - 48$ **[1]** $= 6$ students **[1]** stay in school

20. (a) 2, 7, 12, 17, 22

(b) $5n$ **[1]** $5n - 3$ **[1]**

Find the difference between each term (going up in steps of 5 means $5n$).

(c) When $n = 27$:

$5 \times (27) - 3 = 132$

21. (a) 8 chefs: 140 guests

1 chef: 17.5 guests **[1]**

10 chefs: 175 guests

10 chefs **[1]**

Divide by 8 to find out how many guests one chef can prepare for.

(b) One of the following assumptions:

The same amount of food is prepared for both meals.

Each chef works at the same rate.

It takes the same amount of time to prepare the food for both meals.

22. Volume of ice cubes: $16 \times 2 \times 2 \times 2 = 128\,\text{cm}^3$ **[1]**

Volume of cylinder: $\pi \times 6^2 \times h$ **[1]**

$36\pi h = 128$ **[1]**

$h = \frac{128}{36\pi} = 1.13\,\text{cm}$ **[1]**

23. (a)

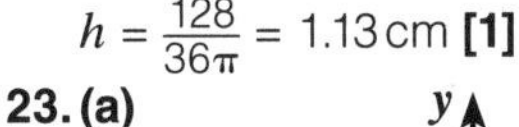

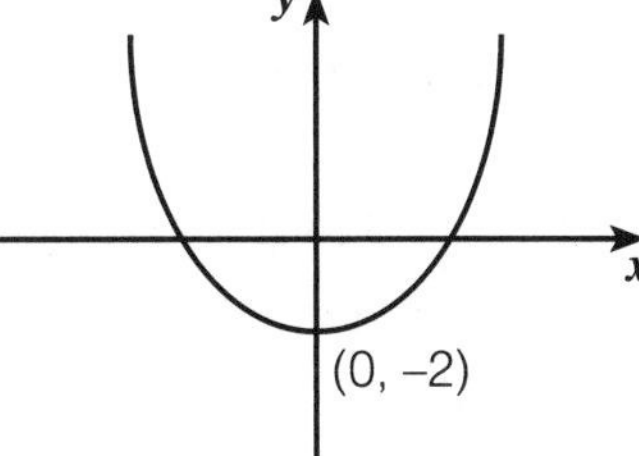

Curve symmetrical about the y-axis **[1]**

y-intercept shown as (0, –2). **[1]**

When $x = 0$, $y = (0)^2 - 2 = -2$, giving the y-intercept as (0, –2). When you square a negative value, you always get a positive answer.

(b) For $y = mx + c$, gradient $m = \frac{\text{change in } y}{\text{change in } x} = \frac{6-2}{5-3}$

$= \frac{4}{2}$ **[1]** $= 2$

Substitute (3, 2) into the equation $y = 2x + c$

$2 = 2 \times 3 + c$ **[1]**

$c = 2 - 6 = -4$

Equation of the line: $y = 2x - 4$ **[1]**

24. $\cos 62° = \frac{\text{adjacent}}{658}$ **[1]**

Adjacent $= 658 \times \cos 62°$ **[1]**

$= 308.91\,\text{m} = 309\,\text{m}$ **[1]**

The angle of elevation from the boat to the man is 62° since alternate angles are equal.

25. (a) $\frac{6^{7-2}}{6^6} = \frac{6^5}{6^6}$ **[1]** $= 6^{-1} = \frac{1}{6}$ **[1]**

(b) $(2^3)^x = 2^4$ **[1]**

$3x = 4$

$x = \frac{4}{3}$ **[1]**

Change 8 and 16 so that they have the same base number, 2.

Paper 3 (pages 62–72)

1. **(a)** $x = 12$ **(b)** $x = 4.5$ **(c)** $25 - 15$ **[1]** $= 10$ **[1]**
2. **(a)** 9 **(b)** 27
3. One square attached to one of the top three squares (or the bottom of the lowest one) **and** one square attached to either side, e.g.

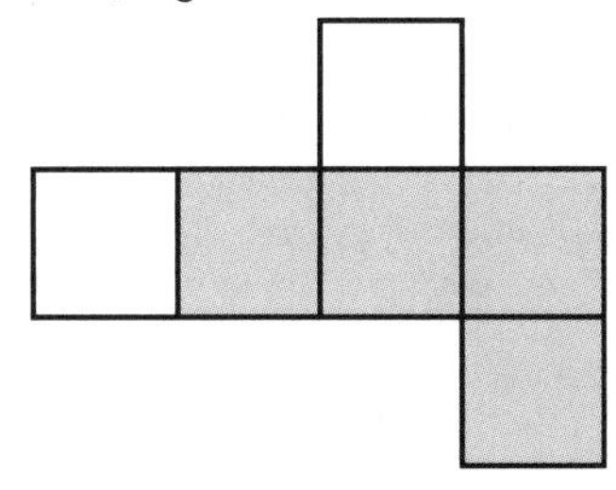

4. **(a)** $-8.3 - 3.2 = -11.5°\text{C}$ **(b)** $-8.3 + 3.6 = -4.7°\text{C}$
5. **(a)** $3 \div 8 = 0.375$

Use your calculator or the bus-stop method.

(b) $\frac{13}{16} - \frac{4}{16}$ **[1]** $= \frac{9}{16}$ **[1]**

(c) $\frac{3}{4} \times \frac{3}{1}$ **[1]** $= \frac{9}{4}$ or $2\frac{1}{4}$ or 2.25 **[1]**

6. $106 + 47 + 29 = 182$ **[1]**

No, because angles in a triangle must add up to 180° **[1]**

7. **(a)** $\frac{3}{9}$ or $\frac{1}{3}$ **(b)** $\frac{2}{8}$ or $\frac{1}{4}$

When Sandeep chooses, there are three white out of a possible nine counters. When Colin chooses, there are only two out of eight counters left which are white.

8. $3(n - 5)$ **[2] [Accept $3n - 15$ or $(n - 5) \times 3$]**

Special case: $n - 5 \times 3$ **[Allow 1 mark]**

9. **(a)** $\frac{3}{7}$ Becky gets three out of a total of seven shares.

(b) $£420 \div (3 + 4)$ **[1]** $= £60$ **[1]** per share

$4 \times £60 = £240$ **[1]**

10. **(a)**

Item	Price	Quantity	Total
Side of salmon	£9.55	1	£9.55
Onions	£0.36	5	**£1.80**
Rice	**£1.18**	2	£2.36
Garlic	£0.17	**3**	£0.51
White wine	£4.49	2	£8.98
		Total	**£23.20**

[3 marks if fully correct; 2 marks with one error; 1 mark for at least one value correct]

(b) 15% of £23.20 $= 0.15 \times 23.2$ **[1]** $= £3.48$ **[1]**

$£23.20 - £3.48 = £19.72$ **[1]**

Alternatively find 85% of £23.20 ($0.85 \times 23.2 = £19.72$)

11. (a)

	Won a medal	Never won a medal	Total
Male	8	**28**	**36**
Female	**11**	**6**	17
Total	**19**	**34**	53

[2 marks if fully correct; 1 mark for at least three correct values]

> Make sure all the columns and rows add up to the correct totals.

(b) 11

12. $2^{-3} = \frac{1}{2^3} = \frac{1}{8}$ or $\sqrt[5]{32} = 2$ **[1]**

$\frac{1}{8} \times 2 = \frac{2}{8} = \frac{1}{4}$ **[1]**

> $\sqrt[5]{32} = 2$ since $2 \times 2 \times 2 \times 2 \times 2 = 32$

13. Berkshire: 115° ÷ 23 = 5° **[1]** per pig

Middle-white: 32 × 5° = 160° **[1]** drawn accurately

British lop: 85° ÷ 5° = 17 **[1]** written in the table

> Measure the Berkshire sector accurately and divide by 23 to find the degrees needed for each pig in the pie chart.

14. An enlargement **[1]** with scale factor $\frac{1}{2}$ **[1]** from the centre of enlargement (6, 4) **[1]**

> Draw projection lines through the corresponding vertices of the two triangles until they all meet at a single point inside the smaller triangle. This is the centre of enlargement.

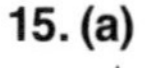

15. (a)

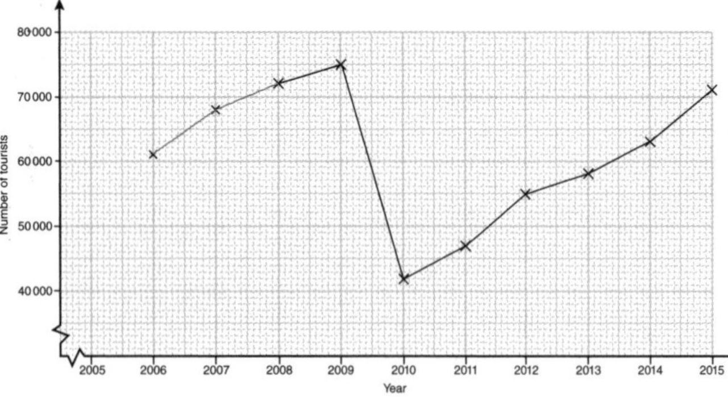

(b) The number of British tourists increased between 2006 and 2009.

There was a large drop in the number of British tourists in 2010.

From 2011 to 2015, the number of British tourists began to rise again.

[1 mark for each correct statement up to a maximum of 2 marks]

16. (a) False: a negative number cubed gives a negative answer (e.g. $(-2)^3 = -8$)

(b) True: a negative or a positive number squared gives a positive answer (e.g. $(-2)^2 = 2^2 = 4$)

(c) False: when $x = 2$, 2^2 is equal to 2×2

17. (a) $-3 \leqslant 2x < 6$ **[1]**

$-1.5 \leqslant x < 3$ **[1]**

> Add 3 to each part, then divide by 2.

(b) –1, 0, 1, 2

18. (a) One of the following reasons:

There is no time period (in a week / month, etc.)

No option for 'never' visit

Two central tick boxes overlap

(b) Add a time period to the original question **[1]**

Have at least four boxes, which include '0' or 'never' and which do not overlap **[1]**

19. (a) $\frac{38}{203}$ of 48 752 ≈ $\frac{40}{200} \times 50000$ **[1]** = 10 000 **[1]** students

> Round each number to 1 significant figure.

(b) An over-estimate since $\frac{40}{200} > \frac{38}{203}$ and 50 000 > 48 752 **[1]**

Assumption: A similar proportion of students from every school in England will not bring a calculator to the exam. **[1]**

20. Premium bonds: £15 × 2.5 × 3 **[1]** = £112.50 **[1]**

Local bank: £2500 × 1.01 × 1.015 × 1.02 **[1]** = £2614.13 **[1]**

The local bank is likely to give the best investment. **[1]**

21. $180 - 8x = 32°$ **[1]**

$x = 18.5°$ **[1]**

Angle $ABC = 180 - 6x$

Angle $ABC = 180 - 6(18.5°)$ **[1]**

Angle $ABC = 69°$ **[1]**

> Use the fact that there are two isosceles triangles, each with an angle sum of 180°.

22. (a) $\frac{10\text{km}}{1.6}$ **[1]** = 6.25 miles **[1]**

6.25 × 133 **[1]** = 831.25 calories **[1]**

> To convert miles to km, 1 mile = 1.6 km

(b) Speed = $\frac{6.25}{1.25}$ **[1]** = 5 mph **[1]**

Answer should be B or D. Since his speed (5 mph) is much slower than his usual average speed, he must be running on difficult terrain. **[1]**

> 1 hour 15 minutes converts to 1.25 hours. Then use the formula: Speed = $\frac{\text{Distance}}{\text{Time}}$

23. Equation 1: $8a - 2b = 14$ **[1]**

Equation 2: $6a - 2b = 9$

Equation 1 – Equation 2: $2a = 5$ **[1]**

$a = 2.5$

Substitute $a = 2.5$ into one of the original equations:

$4 \times 2.5 - b = 7$ **[1]**

$a = 2.5$, $b = 3$ **[1]**

> Multiply the first original equation by 2 to get Equation 1 so that the b terms have the same coefficient.

24.

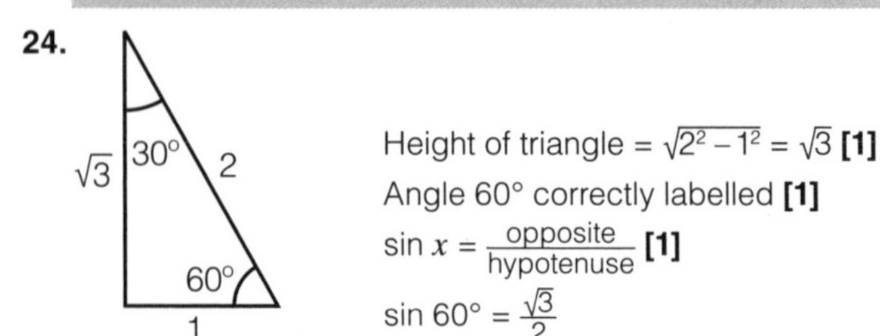

Height of triangle = $\sqrt{2^2 - 1^2} = \sqrt{3}$ **[1]**

Angle 60° correctly labelled **[1]**

$\sin x = \frac{\text{opposite}}{\text{hypotenuse}}$ **[1]**

$\sin 60° = \frac{\sqrt{3}}{2}$

> Use Pythagoras to find the triangle's height ($1^2 + h^2 = 2^2$)